AF787924

FIRST EDITION

FUNDAMENTALS OF REAL TIME SYSTEMS

BY MUKUL SHIRVAIKAR AND THEODORE ELBERT

cognella® | ACADEMIC PUBLISHING

Bassim Hamadeh, CEO and Publisher
Kassie Graves, Director of Acquisitions and Sales
Jamie Giganti, Senior Managing Editor
Miguel Macias, Senior Graphic Designer
Marissa Applegate, Senior Field Acquisitions Editor
Gem Rabanera, Project Editor
Alexa Lucido, Licensing Coordinator
Christian Berk, Associate Production Editor
Joyce Lue, Interior Designer

Printed in the United States of America

ISBN: 978-1-5165-0188-5 (pbk) / 978-1-5165-0189-2 (br)

cognella | ACADEMIC PUBLISHING

CONTENTS

LIST OF FIGURES

DEDICATION

I would like to dedicate this book to my family—my wife, Anjali, and my sons, Vik and Vinny, who supported me throughout the process. I would like to thank Dr. Theo Elbert who introduced me to this area when I was at the University of West Florida. He has the vast experience and knowledge in the areas of real time systems that may be tough to match. I would also like to thank all my colleagues, in industry and academia, who helped me learn this difficult area of expertise, which has morphed rapidly over the last few years. I would like to especially thank Bob Boys and ARM Inc., which helped me redevelop this course after a number of years by providing me with the embedded development tools and support in the form of lab experiments—we truly stand on the shoulders of giants. Finally, I would also like to thank my parents who always encouraged me in my pursuit of knowledge.

Mukul Shirvaikar

To my faithful wife of 62 years, Barbara, who passed away a year before publication.

Theodore Elbert

PREFACE

Real Time Systems provides an introduction to the increasingly complex area of embedded systems as they become commonplace in our daily lives. A real time system is defined as a system that provides a logically correct result in a timely manner as specified by the system requirements. An embedded system is defined as a computer system that is an integral part of a larger system whose primary function is not computational. This book will provide students with an understanding of real time systems concepts, including requirements, analysis, operating systems, software, scheduling algorithms, and, perhaps most importantly, the concept of concurrently executing tasks.

It will introduce students to the tools required to implement successful real world embedded systems projects. The area of real time systems is ubiquitous in the modern engineering world, but the texts available are very applied in nature, or are platform-specific. Some of the texts cover a wide range of material, sacrificing depth for breadth of coverage. The motivation to write this book was to introduce a text that is easy to read and understand for the beginner and one that can be used by the instructor as a complete reference for a course offering.

This book is suited for senior level or graduate level classes on Real Time Systems or Embedded Systems and for students in electrical engineering, computer engineering, and computer science. Students should have taken a prerequisite course in computer programming and, preferably, in microprocessors systems.

A Chapter Overview

Chapter 1 provides an introduction to the general class of "real time" systems. Real time systems may remain in service for many years, and because they may have stringent reliability requirements, it is essential that good systems engineering practice be applied in the development process. Specifically, the use of modularity and abstraction can reduce the complexity of the system, making it more likely to be reliable and easier to maintain or even modify. This chapter covers definitions, characteristics, and an overview of design and implementation issues involved with real time systems.

Chapter 2 provides an introduction to the host-target development environment for embedded systems, with an emphasis on the C programming language. This coverage includes specific constructs normally found in the development of multitasking, real time software. The intent is to cover only the major topics of the C language without going into details that can be found in any reference text dedicated to the subject.

Chapter 3 covers the concept of "structures" using the C programming language as it is commonly used and acts as a natural base leading to objects and object-oriented programming. Implementation examples with a real time operating system are also provided where appropriate. This chapter covers the various design approaches implemented by system designers to meet real time requirements. In particular, the difficult problem of scheduling concurrently executing and communicating tasks is introduced. The three general classes of scheduling are ad-hoc methods, deterministic methods using a repetitive structure commonly termed a cyclic executive, and non-deterministic, priority-driven scheduling using a multitasking executive. Examples of these methods will be presented in later chapters.

Chapter 4 introduces the reader to the concept of a real time operating systems. The principal characteristic of most real time systems is the concept of concurrently executing tasks, which typically communicate with one another. The various scheduling techniques to be presented later are introduced here. This chapter also introduces the concept of scheduling concurrently executing tasks. From basic ad-hoc scheduling to the more sophisticated interrupt-based methods, to the use of a cyclic executive or Real-Time Operating System (RTOS), the various task scheduling methodologies are introduced, serving as a primer for the detailed descriptions presented in **Chapter 5**.

Chapter 5 covers a variety of real time scheduling algorithms in detail, starting with aperiodic tasks and moving on to periodic tasks on multiple processors. Deadline-monotonic, rate-monotonic, earliest deadline, and least slack time scheduling algorithms are covered for single and multiple processors.

Chapter 6 covers the various communication mechanisms typically available in real time systems, including mailboxes, semaphores, and signals. The chapter then discusses some reliability protocols that have been established over the years to prevent system failure, including the priority inheritance and the priority ceiling protocols.

Chapter 7 discusses some additional topics and techniques that are useful while designing real time system software, including cyclic executive design. The topic of handling aperiodic tasks is, then, addressed in a formal fashion.

It will be advantageous to anyone using this textbook for a course to supplement it with some experiments on a real system. A number of such systems are available now and are very economical compared to systems in the past. The author uses the STMicroelectronics Discovery Board with experiments on the RTX operating system with ARM-Keil MDK IDE tools, courtesy ARM Inc. The students thoroughly enjoy the break from theoretical scheduling problems and can actually see them in practice, enhancing their learning experience and providing them valuable experience for their careers.

CH 1

INTRODUCTION TO REAL TIME SYSTEMS

The advent of microprocessors and microcomputers, and their subsequent miniaturization and cost reduction over the last few decades, has resulted in their ubiquitous presence in almost every product used in modern, everyday life. Microcontroller systems may be stand-alone or embedded in products, and they may include specialized hardware and software to meet performance requirements. This chapter provides an introduction to the general class of *real time* systems.

1.1 Introductory Concepts and Definitions

The term *real time system* usually refers to the embedded computer and the hardware and software components of an operational system. Typically, the software component of the system has the most real time constraints placed on it and is written as a program that interacts with the hardware. Software performance can be one of the most critical issues when characterizing a real time system. Although the operation of hardware, along with its environmental requirements, may be important in any given case, the fact remains that the computer chips and peripherals employed in real time applications are not too distinctive from those employed in more conventional applications. Real time computers or applications differentiate themselves from conventional personal computers, engineering workstations, or mainframe machines based on the software running on them and their careful design. One of the most distinguishing features of real time software is the element of *response time*. The software must guarantee that time deadlines established by the requirements of the system are met consistently, even repeatedly, at specific periodic rates as necessary. Failure to meet these timing requirements may constitute an error as serious as, or, perhaps more serious than, the calculation of an incorrect result.

Definition

> *A real time system is defined as a system that provides a logically correct result in a timely manner as specified by the system requirements.*

As is evident from the definition, the system requirements can play a significant role in determining the performance of a real time system.

The claim can be made that all computer systems are real time systems in some sense. There are differences in consequences for missed deadlines, depending on the specific functionality of the system. Every computing system has some response constraints since turnaround time is important to users and has economic and other ramifications. Even in interactive web-based systems, as found on the Internet today, response time is important to users, and excessive response time leads to inefficiency and may result in failure due to customer satisfaction issues.

Some examples of real time systems are:

- Factory Inspection System
- Internet Search Program
- Digital Camera
- Car Antilock Braking System
- Microwave Oven
- Computerized Exercise Bike

The last three systems represent a special subset of real time systems known as embedded systems because they are computer controlled systems that have to function on a stand-alone basis supporting tasks that are not strictly computational in nature. Figure 1.1 shows several real time systems based on their level of complexity.

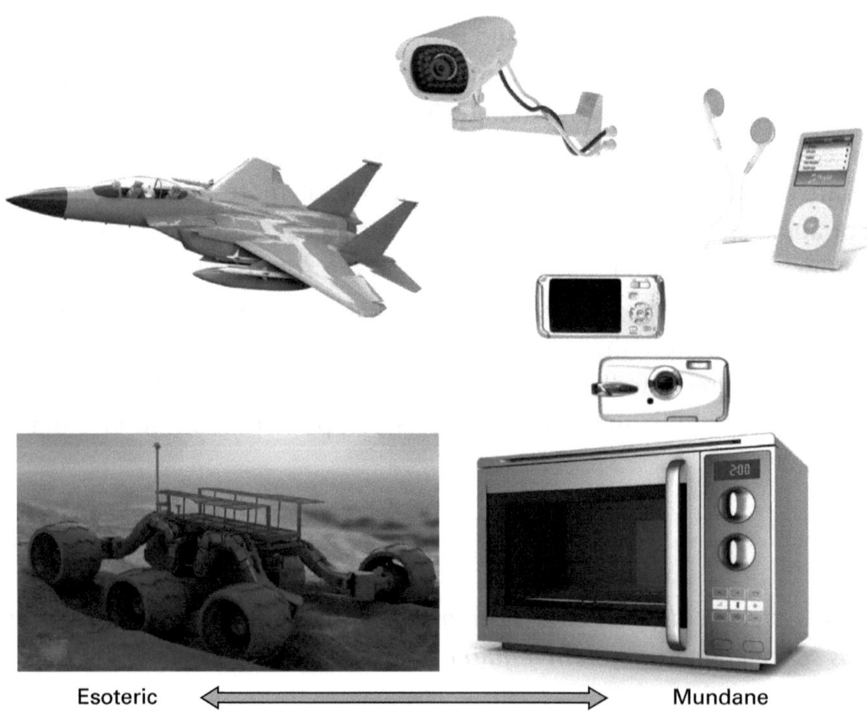

Figure 1.1 Typical Real Time Systems Found in Everyday Life.

Definition

An embedded system is defined as a computer system that is an integral part of a larger system whose primary function is not computational.

The embedded computer is a component that performs a computational task supporting the overall functionality of the system. It is in the area of embedded systems that software performance becomes critical to success. In such systems, the computer must interact with external processes extremely quickly, in milliseconds, microseconds, or even at the nanosecond level. This software-hardware interface typically consists of monitoring an external process through measurements and, then, issuing directives to control the physical behavior of the system based on these measurements. The observation and control functions both, typically, have to meet time constraints, and the burden of meeting these time constraints is almost always placed on the software, and thereby, on the software developer.

The performance requirements placed on real time systems vary significantly based on the application and may differ by orders of magnitude. Interactive real time systems that primarily work with human operators and real time systems with more stringent specifications possess the following distinguishing features.

Interactive systems: Such systems typically react with human operators, and the required response time requirements are in the order of seconds. The response time requirements are typically based on the ability of the human operator to recognize and react to data generated and presented by the computer.

Real time systems: Such systems typically react with other hardware subsystems with transducers, and the response time requirements are on the order of milliseconds or even microseconds, dictated by the rate of observation and by the rapidity at which the external environment can change. The function of the real time system is to provide the required interaction with the external environment when the required rate of interaction, in most cases, transcends the ability of a human operator. A general real time system can be described as shown in Figure 1.2. The components of the system, in addition to the embedded computer, are sensors, which observe or measure the appropriate attributes or variables of the real world process, and actuators, which are used to exert influence on the process. Additionally, interfaces are necessary to convert sensor readings to a digital format

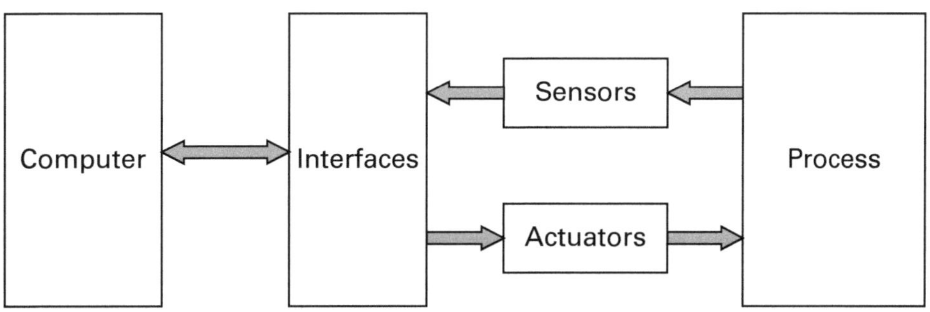

Figure 1.2 A Typical Block Diagram for a Real Time System.

acceptable to the computer and to translate computer-generated commands to signals that drive the actuators.

For example, consider a typical real time system—an aircraft autopilot whose purpose is to automatically control the aircraft heading and altitude. The real world process, in this case, is the aircraft, with sensors providing the direction, the altimeter, and inertial instruments that provide information about the attitude of the aircraft and the rate of change of attitude. The actuators are the electrical or hydraulic controls that move the control surfaces in response to commands generated by the embedded computer. The aircraft has a dynamic response to control inputs which is fed back to the system in a closed loop, thereby forming a real time control system.

Some definitions may help in the subsequent discussion of real time systems and, in particular, of the software that drives the associated embedded computer.

Definition

Response time is defined as the time from a request for service or excitation to the time when the service is provided or requisite system output is presented by the system.

The real time response is the completion of the necessary operations, possibly periodic in nature, including computation and input/output operations when one or more stimuli or triggers are presented to it. Such triggers may include timers that permit the system to be successful in keeping up with the passage of real world time deadlines. These timing requirements usually are expressed as *deadlines*—time increments following a request for service during which the service must be provided. As mentioned, requests for service often are periodic in nature, but they may also be aperiodic, triggered by events occurring in the real world process.

Definition

A real time embedded computer is an embedded computer that must provide the necessary real time response for proper functioning of the real time system.

The ability of real time embedded computers to meet the real time response requirements imposed upon them depends on the real time software, assuming the hardware has been chosen by the system designers to meet processing requirements.

Definition

Real time software is software that executes on a real time embedded computer.

Real time software development refers to the development of software for real time embedded computers. Because of the stringent real time response requirements that must be met, the development of real time software is different from that of more conventional software. Principally, the additional requirement for proper timing is as important as the correctness of the results produced. There is a wide range of real time systems, and they have different system specifications or requirements in terms of timeliness. One of the popular approaches is to classify a real time system based on the criticality of meeting deadlines:

- *Hard real time system*—one in which missing even a single deadline will result in failure or catastrophe.
- *Firm real time system*—one in which a deadline may be missed very infrequently in order for it to be considered as successfully operating.
- *Soft real time system*—one in which deadlines may be missed frequently without major adverse effects on the system behavior.

Due to the wide spectrum of systems, it can be difficult to classify them as one of the above. Instead, it may be preferable to subdivide the system into parts, with different levels of requirement—hard, firm, or soft.

This book is about real time systems development and, as such, the discussion will concentrate on those concepts, methods, and techniques unique to the development of real time systems. In some instances—notably, when examining methodologies for the specification and design of real time software—extensions to the more common techniques used in the development of conventional software will be described.

1.2 Real Time System Features

The range of characteristics exhibited by real time systems is quite broad since they span a wide range of applications from microwave ovens to space stations. Most applications of interest, however, tend to exhibit the following characteristics:

- *Large software component.* This attribute definitely isn't unique to real time systems and is a direct result of the rapid growth of hardware capability of computers since their invention and rapid adoption in daily processes. In an endless cycle, more capable hardware results in greater expectations from the system, and with this greater expectation, comes larger and more complex software. Today, tens of thousands of lines of code is common, hundreds of thousands of lines of code is not uncommon, and many systems have software components with millions of lines of code. The complexity of systems software grows at a non-linear rate with size. Since complexity presents the biggest obstacle to the development of well-engineered software, the use of sound software engineering practice is imperative in real time systems software development. This is even more critical when the additional complexity of the timing constraints inherent to real time software are considered. Most software specification and design methodologies do not consider timing constraints. However, they are important to the systematic design process characteristic of an effective engineering approach to software development. Real time extensions to these methodologies have been developed, and some of these are discussed in future chapters.
- *Long life.* Operational real time systems tend to be in service for long periods of time. Defense weapon systems have effective lives in excess of twenty years. Some spacecraft journeys last for decades, and it is typical for the software onboard to be modified during the life of the vehicle.

Applications that manage the drive trains in automobiles are used for extended periods of time, with modifications constituting the primary change from one model year to the next.

- *Continuous change in requirements.* Real time systems remain in the field for extended periods of time. Consequently, it is common that they undergo modification due to changing operational requirements or require maintenance updates to correct software errors. Today, it is common to design systems so that operational modifications can be implemented as software changes rather than hardware changes. This feature further underscores the importance of using sound software engineering practice in the development of real time systems software, to enhance maintainability.
- *Physical constraints.* The physical environment imposed by the system operating conditions may limit the hardware size and power consumption. These limitations can impose constraints on the software, particularly in the presence of real time scheduling requirements.
- *Specialized processing operations.* While conventional computers provide input/output operations in a form adapted to textual or graphic data for use by printers, keyboards, terminal screens, etc., real time systems must deal with sensor data, actuator commands, analog-to-digital and digital-to-analog conversion, and other forms of information.
- *Concurrent processing or multitasking.* The timing constraints placed on the software often relate to multiple facets of the application domain. There may be various sensors and actuators, each of which imposes a timing constraint. The evolution of the physical processes themselves is characterized by temporal features, which may impose time deadlines on the software functions. The time-tested method in which a single processor can meet such a diverse set of timing constraints is through the use of concurrent processing. The fundamental unit of concurrent processing software is the *task*—a logically complete section of code that can be independently scheduled for execution. Tasks are typically designed to address one or more modular processing requirements dictated by the system specifications. Often, such a requirement is periodic in nature and results in a task that executes repetitively at a specified rate. Such tasks are termed *periodic tasks*. Tasks that do not have a periodic requirement are called *aperiodic tasks*. It is quite common for a given system to have both periodic and aperiodic requirements.

These general characteristics of real time systems, together with the timing constraints imposed by the associated physical processes, lead to general software requirements that, in some instances, differ from those of conventional software. In particular, concepts, such as reliability, maintainability, response to unexpected situations, and timely response to external stimuli, take on new and critically important roles. Figure 1.3 shows the expanse of real time embedded systems encountered in a typical modern aircraft today. The following section addresses the general requirements of real time software and stresses those aspects of software engineering that are particularly applicable.

1.3 Real Time System Requirements

While in the early decades of building real time systems for controlling physical processes, it was common for the hardware to be much more expensive than the software; today, this position has

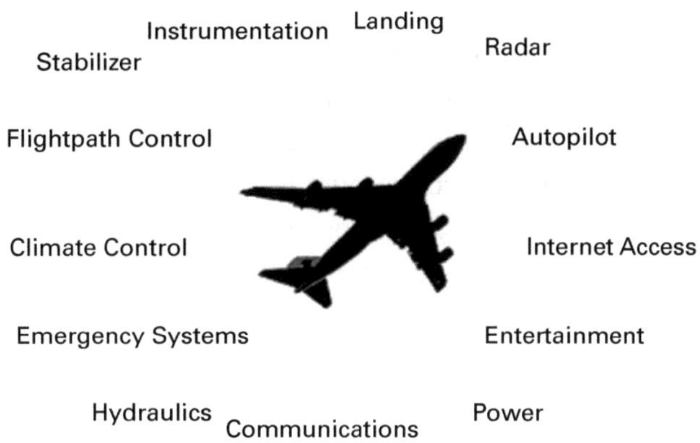

Figure 1.3 Embedded Systems in a Modern Aircraft.

been reversed significantly. By far, software is usually the most costly component of real time embedded systems. Not only does it take more time to develop but software is also, remarkably, the most unreliable element of the system due, principally, to the infinite possible situations with which it can be presented. It is a matter of published record that software errors have resulted in a number of mission or system failures. In response to this scenario, a set of general requirements for real time systems software has evolved to reflect the critical role it plays in overall system reliability and because more emphasis is being placed on systems software as the means for addressing changing operational requirements and application versatility. These general requirements or attributes that most real time systems software and hence, the systems themselves must exhibit (not to be confused with specific requirements as determined in the requirements analysis phase of product development) are presented in the following list.

- *Timeliness.* The correct result delivered late can constitute a system error, and from the user or application viewpoint, the response to external stimuli must often appear to be instantaneous or nearly so. The system must also be adaptable in responding to external stimuli since the sequence and timing of external, and sometimes, internal stimuli, are not in the control of the system developer. Particularly, while the scheduling of periodic tasks is synchronous with respect to the internal clock of the embedded system, those tasks responding to external stimuli are usually asynchronous in nature.
- *Correctness.* The system must always produce correct results in order to prevent a failure or catastrophe, and this rather obvious characteristic is referred to as *correctness*. In practice, correctness is very difficult to prove for systems of any appreciable complexity due to the infinite possible permutations and combinations of states and inputs. The technique most used in attempts to demonstrate the correctness of software is testing, and for most systems, this is incomplete since only a small subset of scenarios can be considered. There has been some modicum of success in applying tools from set theory and predicate calculus to the proof of correctness of systems, but these formal methods have never matured to the point that they can be used to prove correctness of large, complex systems.
- *Versatility.* The fact of the matter is that most errors in real time systems arise from situations that were not encountered in testing. The system must be developed such that it can respond

correctly to new situations and even recover from error situations, if necessary. This trait is termed *completeness or versatility*. For example, exception handling is an important feature of most real time software that permits recovery from unusual and/or unexpected situations.

- *Reliability.* Real time systems, in particular, must not fail because of the simple fact that such failure can produce catastrophic results, perhaps, even the loss of human life.
- *Maintainability.* Products and systems tend to remain operational for extended periods of time, and changes in the field to respond to operational requirements are common. Real time systems must be designed with maintenance in mind. Strong modular design and a systematic design technique with unit testing and other approaches are commonly used for enhancing system maintainability.

In addition to the above characteristics, the attribute known as *reusability* is very beneficial to systems development of any kind. Software and hardware reuse refers to the use of components in multiple systems. The idea is relatively new for software components for obvious reasons (hardware such as microprocessor chips have been modular for a while). If certain software functionality is required in a program, the software developer simply goes to a catalog and selects a component to provide that functionality and incorporates it into the program. While this concept seems rather straightforward, and programmers have, for decades, used libraries of subroutines to avoid rewriting the same code over and over again, many practical problems relating to differing languages, various platforms, and operating system vagaries have complicated the issue. With modern, platform independent languages like Java, it is possible to build a truly reusable software component ("Write once, use many times" is the Java motto). The advantage of software reuse is multifold: reduced development costs since the software does not have to be developed and inherently accumulated reliability since a software component obtained from a reused library most likely has accumulated some operational use and has been deployed in the field.

1.4 Design and Implementation

The design and implementation of real time systems involves special constraints, including cost and performance. While the former is driven by market forces, the latter has a number of aspects that are application-specific, such as size, weight, power, energy, environment, maintainability, and reliability. Engineers designing real time systems must balance all these factors to make judicious choices for the hardware and software implementation for the system. It is extremely important for the engineering team to carefully take input from the end user or customer and consider the implications of their design choices on the product. This is typically done using a system specification document, which can be very valuable as a reference for all parties involved.

Figure 1.4 shows the typical hardware options for building a real time system along with the implementation tradeoffs. The options can be classified into two general classes:

- *Hardware Intensive Design.* This approach typically involves the design using discrete fixed-function logic, Application Specific Integrated Circuits (ASIC), Programmable Logic Devices (PLD), and Field-Programmable Gate Arrays (FPGA). ASICs are chips designed for specific applications

Real Time Design Solution			
Hardware Intensive Designs		Software Intensive Designs	
System-on-a-Chip, ASICs	FPGA, CPLD, PLA, PLC	Microprocessors, Microcontrollers, Daughter Boards	Turnkey Systems, Embedded PC
High Development Cost	Medium Development Cost	Low Development Cost	Medium Development Cost
Very High Speed	High Speed	Medium Speed	Medium Speed
Very difficult to modify/upgrade	Moderately difficult to modify/upgrade	Moderately difficult to modify/upgrade	Very easy to modify/upgrade
High volume applications	Medium/high volume applications	Medium/high volume applications	Low volume applications
Low unit cost	Medium unit cost	Medium unit cost	High unit cost
Slow time-to-market	Medium time-to-market	Medium time-to-market	Medium/low time-to-market
Small size	Small size	Medium size	Large size
Low power	Low power	Medium power	High power

Note: the above are general observations and may not be true for every real-time System, as the range and application variability is very wide.

Figure 1.4 Options for Building Real Time Systems.

and cannot be changed once they are manufactured. PLDs and FPGAs, on the other hand, have logic circuit blocks that can be reconfigured on the fly using special programming languages by the designers. The implementation team focuses on creating custom hardware solutions to obtain higher operational speed with less focus on software.

- *Software Intensive Design.* This approach typically involves the selection of hardware available from other suppliers or manufacturers such as, but not limited to, microprocessors, microcontrollers, or embedded PCs. The implementation team focuses on creating custom software solutions to obtain faster time-to-market with less focus on hardware.

Various combinations of the above approaches may be used by system designers based on the requirements of the application and other factors such as development cost and targeted time-to-market requirements for the product.

Due to various considerations, such as cost, training, and time-to-market, software development is not always implemented as the mature engineering discipline that is required for reliable real time system development. Fortunately, progress is being made due to the development of the formal discipline known as software engineering which is defined as the application of a systematic, disciplined, and quantifiable approach to the development, operation, and maintenance of software. Software engineering is concerned with the entire life cycle of software and, thus, addresses the operational phase, in which maintenance and modification take place, as well as the development phase. One of the principal axioms of software engineering is that the complexity of software, and, particularly, the complexity inherent in large software systems, is to blame for most of the problems relating to reliability and maintainability. The software engineering concepts that hold the most hope for managing the complexity of software are modularization and abstraction, and the analogy to the design of hardware and total system development cannot be overlooked here. Modularity (i.e., the use of functional modules as building blocks and the description of these modules in terms of functional behavior rather than method of operation) is the primary means the hardware designer uses to manage the overall complexity of the design. Reliability and maintainability are also enhanced by such a modular approach.

It is necessary to impose a strong modular structure during the design phase of the development cycle. This effectively results in the creation of modules that are cohesive and have defined and minimal mutual interaction. When modules do interact, they should do so only through well-defined interfaces that are under control of the developer. A strong modular structure incorporated in the design phase reduces the complexity of the software and makes system development more manageable. It has also been shown to reduce errors in the resulting software and increases in reliability and maintainability. The use of modularity in the design of real time systems strongly corresponds to the general characteristics of real time software requirements. Specifically:

- Multiple methods of interaction with the world.
- System specifications and functionality that may be interdependent.
- Synchronous or periodic functional requirements.
- Asynchronous or aperiodic functional requirements.
- Strict response times or timing tolerances.

As noted earlier, the requirements imposed by the above characteristics are best serviced by concurrent processing, and fundamental to the concept of concurrent processing is the *task*. A task is considered to be a processor, the program code, memory, and input/output resources necessary to perform a particular function. In order for a task to execute, the necessary resources must be available, and if they are not available, the task execution must be postponed. Since the task is the fundamental element of concurrent processing, and since concurrent processing is necessary to meet the timing constraints imposed on real time software, it seems reasonable that the program code of the tasks be the primary unit of modularization. In practice, such is actually the case, with further modularization within tasks provided by subprograms or by other means.

Modularization into tasks is based on assigned functionality since each task performs a specific function. Multiple functions may be required sequentially, simultaneously, or in some combination thereof. Furthermore, the system functional requirements may be synchronous or asynchronous. In other words, the requirement can be scheduled by the internal clock (synchronous), or it may arise randomly as a result of external events (asynchronous). Accordingly, the tasks may display different possible behaviors:

> Tasks may execute in sequence with each task running to completion in a deterministic scheduling system; or
>
> Tasks may execute concurrently through the use of non-deterministic scheduling.
>
> In either case they will be interrupted by asynchronous events that cause the execution of aperiodic tasks.

There are three general approaches to task scheduling, namely, ad-hoc scheduling; deterministic scheduling using a cyclic executive; and non-deterministic, priority-driven scheduling using a multitasking executive.

- *Ad-hoc scheduling.* This is the simplest form of task scheduling if it can be referred to as scheduling at all. In ad-hoc scheduling, the tasks are functional program units, and intertask dependencies are limited to precedence relationships. A precedence relationship is simply a requirement that one task complete prior to the execution of another. The scheduling of the tasks is merely an arrangement whereby a currently executing task starts the subsequent task after its own completion. This procedure continues until all tasks have run to completion. When this happens, the first task is executed again, and the process repeats. Task execution is repetitive in a cycle, but it is not necessarily periodic. This can be attributed to the fact that the repetitive process does not occur at a precise predictable rate, but rather, a task repeats whenever its predecessor task completes execution. Therefore, no specific periodic requirements can be specified for task execution. The program is essentially written as an endless loop with each task executing to completion in some predetermined sequence and with asynchronous events handled by interrupt processing. Since this form of task scheduling is rather straightforward and of limited use, it will not be covered in too much detail.
- *Cyclic scheduling.* In cyclic scheduling, tasks are deterministically scheduled on a periodic basis by an interrupt-driven program called a *cyclic executive*. The cyclic executive obtains its timing "clock ticks" from a sequence of interrupts generated by a timer, which must be an integral part of the embedded system hardware (typically the microcontroller). Task scheduling in a cyclic executive system is described better by the term *periodic* rather than repetitive. In essence, the tasks are scheduled at a precise periodic rate. Asynchronous events are handled by interrupt processing or by background processing or by special periodic server tasks. The processing required by the occurrence of asynchronous events is, by definition, non-deterministic. Otherwise, the scheduling of tasks in a cyclic executive system is completely deterministic.
- *Priority-driven scheduling.* In priority-driven scheduling, tasks are assigned priorities based on a predetermined strategy. They are, then, dynamically scheduled based on the priority assignments. At any point of time, the task with the highest priority will be scheduled for execution unless its execution is blocked. A task can be blocked if the current state of the program determines that it should not be allowed to execute. For example, the task may be waiting for another task to provide it with needed data, possibly with a different priority. Priority-driven scheduling is provided by a high-priority task or software system component called a scheduler. A multitasking executive or operating system usually provides such a scheduler service. Additionally, such powerful executives usually provide other services such as task synchronization, intertask communication, and memory management. Processing required by asynchronous events is simply designated to be another task that is assigned an appropriate priority and scheduled accordingly by the multitasking executive. All task scheduling is non-deterministic since the tasks are dynamically scheduled as the program executes.

The ad-hoc approach is extremely useful for simple systems that need to be designed under extreme cost and speed constraints. It should not be discounted because for rather straightforward processing with no specific periodic requirements, it can provide satisfactory results. When used with interrupts it can also provide the processing associated with external asynchronous events. However, when there is a requirement for strict periodicity, cyclic executive or priority-driven scheduling must be used. The cyclic executive approach provides deterministic scheduling and low implementation overhead since the cyclic executive software is small and fast. The priority-driven approach provides increased flexibility and power since tasks can be added or deleted, or the priorities of tasks can be changed without necessity of redesign. The cyclic executive is appropriate for systems that have these characteristics:

- synchronous (high periodic content)
- predictable (tasks with predictable execution times)
- relatively few asynchronous requirements
- deterministic execution sequence

An excellent example of a cyclic executive system is an inertial navigator for an aircraft: the inertial instruments provide data at precise periodic rates; data must be processed prior to the arrival of the following data; asynchronous events, including operator control panel entries, are relatively rare; and task execution times are fairly predictable since they are symmetric and involve approximately the same amount of calculation each time they are executed.

On the other hand, the multitasking priority-driven approach is appropriate for systems that have the following characteristics:

- synchronous deadlines
- non-deterministic tasks (execution times of the tasks may vary over successive execution instances)
- large number of asynchronous events

An example is a C^3 (command, communication, and control) system in which there are some periodic tasks but which must respond to large numbers of asynchronous events as data are collected and processed by the system. Real world systems typically lie somewhere between these two descriptions, often making the design choice of what approach to take a difficult one.

1.5 Summary

Real time systems and their design and development are significantly different from that of more conventional systems, principally, because of the importance of timing constraints or deadlines in addition to normal reliability and correctness requirements.

- Timing constraints may be specified as periods or deadlines.
- The timeliness requirements make missing deadlines a serious matter (possibly more serious than computational errors).

- A set of tasks is typically used as the solution for meeting time deadlines.
- Task scheduling has to be utilized to meet system specifications.

Other than ad-hoc task scheduling, the tested methods for meeting system requirements are the timer-driven cyclic executive approach and the priority-driven scheduling using a multitasking executive. The cyclic executive approach has the following characteristics.

- Simple implementation.
- Low overhead in both code and execution time.
- Deterministic scheduling.
- Inflexible with respect to adding or deleting tasks.
- Intolerant of variations in execution times of the tasks.
- Difficult to modify since only minor modifications can cause severe timing problems.

On the other hand, software built around a multitasking executive exhibits almost an opposite set of characteristics.

- Complex implementation.
- High overhead in both code and execution time.
- Non-deterministic scheduling.
- Flexible for adding or deleting tasks because scheduling is dynamic.
- Tolerant of task execution times variances.
- Easy to modify since minor modifications do not usually cause timing problems.

Task synchronization, which is built into the schedule of a cyclic executive, must be provided dynamically in a priority-driven system. Multitasking executives provide the means of dynamic synchronization and communication through the use of semaphore and event queues.

Real time systems may remain in service for many years and may have stringent reliability requirements. It is of prime importance that good systems engineering practice be applied in the development process. Specifically, the modularity and abstraction can reduce the complexity of the system, making it more likely to be reliable and easier to modify. The remainder of this text delves into the various methods of analysis and design of real time systems, specifically focusing on the multitasking executive and task scheduling approaches that are becoming cost-effective and prevalent in an increasing number of applications.

Image Credits

CH 2

REAL TIME SYSTEMS PROGRAMMING

Real time systems programming poses some unique challenges, as outlined in the introduction, due to the demanding requirements imposed on embedded systems. The primary requirement is reliability, which includes the ability to meet deadlines under harsh operational conditions. This chapter provides an introduction to the software development environment for such systems with an emphasis on the C programming language. Included are specific constructs used in the control and synchronization of concurrently executing tasks.

2.1 Host-Target Development Environment

One of the unique features of real time systems is that a large number of them are embedded in nature—they are "stand-alone" and have to operate on the underlying hardware with minimal or no connection to the outside world. The underlying hardware is typically a microcontroller board and may not be complex or powerful enough to have compilation and linker capability. It is capable of running an application program loaded in memory, typically known as *firmware*, but cannot be used for any software development functionality other than testing. As a result, there are typically two modes for such systems—the development or test mode, and the operation mode. In the development mode, the underlying hardware or microcontroller is tethered to a more powerful system such as a networked computer.

The former is known as the *target* system and the latter is known as the *host* system. The software is developed on the host machine but is developed to execute on the target machine. In more sophisticated environments, the host machine may contain a target machine emulator. In this case, the software executes on the emulator within the host machine during development and on the target machine when deployed. Such an arrangement is typically known as a *host-target development environment*. Figure 2.1 shows a typical configuration for such an environment.

The host-target development environment is used by the system developers to create and test the application software program. It can also be used to test the various hardware components that are part of the system—memory, sensors, and peripherals. Such an environment can be quite challenging to work in because of the timing complexities and observability inside the target. Consequently, a lot of effort has gone into special tools that help the real time engineers to accomplish their tasks. The goal of the real time system engineers is to ensure that the target can operate independently with an

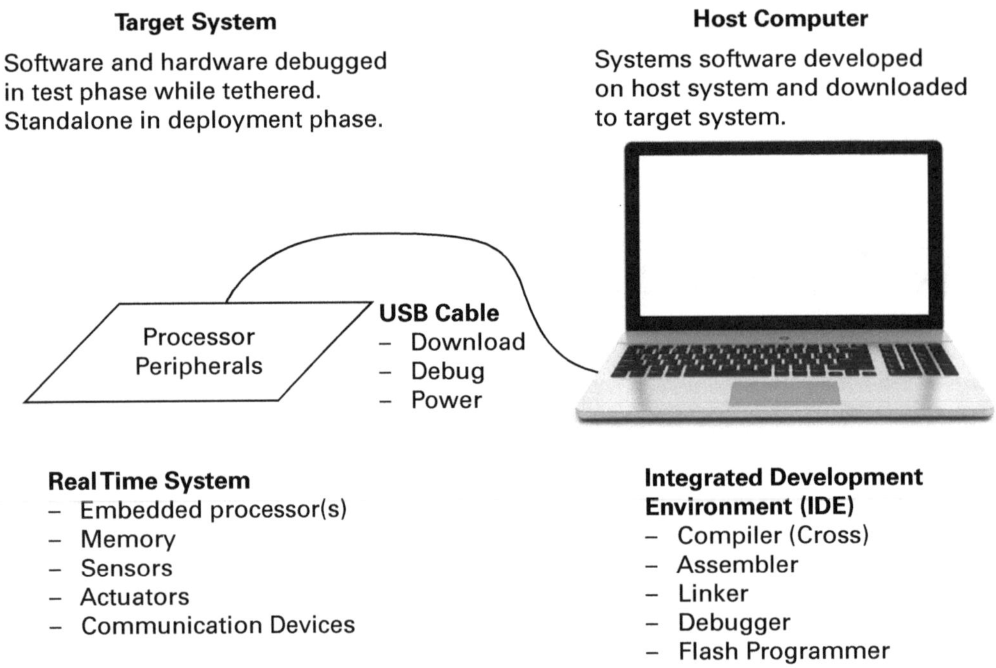

Figure 2.1 The Host-Target Development Environment.

acceptable level of reliability (typically, this level is determined by the system requirements). Once this is achieved, the target can be deployed to the field and will be in the operation mode.

2.2 The C Programming Language

The C programming language is still the most popular language of implementation for real time and embedded systems. The primary reason for this fact is based on the low-level nature of the language and the ease of interfacing it with the underlying hardware or the *instruction set architecture* of the processor on which the code is actually executed. The C language was originally developed in the 1970s by Ken Thompson and Dennis Ritchie for the UNIX operating system. Thompson and Ritchie won the Turing Award "for their development of generic operating systems theory and specifically for the implementation of the UNIX operating system." Linux is also written in C. Some of the characteristics of C that make it suitable for real time systems are:

- Small code size
- Emphasis on function calls
- Loose typing (mixed-mode arithmetic allowed)
- Structured programming (if, for, while structures)
- Allows bit-level manipulations
- Extensive use of pointers

Most real time development today is performed on an Integrated Development Environment (IDE). These environments typically include a graphical user interface that allows the developer to create projects, include libraries, and provide a hardware interface for specific platforms. They also include a *software development toolchain* that is customizable by the developer. Most modern IDEs also allow the user to perform some level of real time debugging of their code on the target, with the help of tools such as code tracing one line at a time, setting breakpoints and watch points, and using variable and memory watch windows. The software toolchain typically consists of compilers, assemblers, and linkers provided by the vendors or chip manufacturers. A typical software development toolchain is shown in Figure 2.2.

The various tools that can be found in an IDE and their functions are as follows:

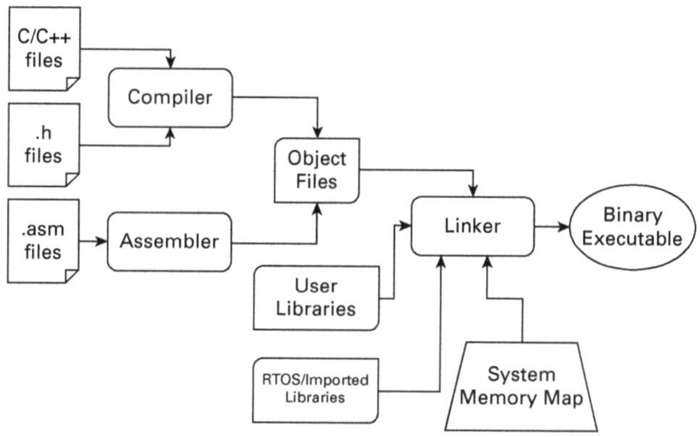

Figure 2.2 Software Development Toolchain.

- **C Compiler**—converts C programs into object code files
- **Assembler**—converts assembly code files into object code files
- **Linker**—joins multiple object files together into an executable file as per memory configuration for execution on the target
- **Flash programmer—**programs the executable file into the flash memory of the target
- **Debugger**—interactively controls the operation of the target processor, allowing the developer to view target memory and program information during development operations
- **Simulator**—allows the program to run on a simulator without real hardware
- **Utilities**—tools such as file converters and data managers for trace purposes

Some examples of IDEs are the Keil MDK-ARM Microvision by ARM Inc. and Code Composer Studio by Texas Instruments Inc.

2.3 Data Types

The data types in any programming language correspond to the definitions of different types of data that are stored, displayed, and manipulated by the computer for an application. Typically, these data types will include formal definitions to represent bits and Boolean numbers, integers and real numbers, characters and strings, computer addresses and collections thereof. Not all data types supported by a language may be supported by the hardware processor on which the program runs.

Data types allow the declaration of variables and constants to reserve or allocate memory storage on the computer. The basic C programming data types are the following:

- *char*: −128 to 127
- *short int*: −32,768 to 32,767
- *long int*: −2,147,483,648 to 2,147,483,647
- *unsigned short int*: 0 to 65,535
- *int*: are assigned as either *short int* or *long int* depending on the computer
- *float*: "single precision"
- *double*: "double precision"
- *long double*: rarely used

Table 2.1 Data Types in C and Common Types Used in Real Time Systems.

C Data Types	Bits	Signed Range	Unsigned Range
char, int8_t, uint8_t	8	−128 to 127	0 to 255
short, int16_t, uint16_t	16	−32,768 to 32,767	0 to 65,535
int, int32_t, uint32_t	32	−2,147,483,648 to 2,147,483,647	0 to 4,294,967,295
Long	32	−2,147,483,648 to 2,147,483,647	0 to 4,294,967,295
long long, int64_t, uint64_t	64	−9,223,372,036,854,775,808 to 9,223,372,036,854,775,807	0 to 18,446,744,073,709,551,615
float	32	3.4E +/− 38	
Double	64	1.7E +/− 308	
long double	64	1.7E +/− 308	
Pointers	32	0 to memory width	
Enum	8, 16, 32		Smallest possible
bool (C++), _Bool(C)	8		false or true

C allows mixed mode arithmetic, i.e., while writing a program, different types of variables can be used within a single statement, and the compiler will convert the variables to the appropriate type using a set of rules:

- Variables of type *char* are always converted to *int* before an operation is performed.
- Operations of variables of the same type always produce results of that same type.
- When variables of different types are found in an expression, variables of a *lower* type are promoted to the *higher* type (i.e., an *int* added to a *double* would get promoted to *double*).

This usually works fine but can produce some interesting results if not dealt with carefully:

Example 2.3.1: Consider the C code below that shows an example of *typecasting.*

```
Double ratio;
int x, y;
```

```
x = 5;
y = 2;
ratio = x / y;
```

x / y would be evaluated using integer division, result would be ratio = 2.0.

```
double ratio;
int x, y;

x = 5;
y = 2;
ratio = (double) x / y;
```

Variable x is converted to type double, y is also promoted to double, and the result is computed as double. This is an example of *typecasting* in C.

C also has an extremely useful construct known as an enumeration. An enumeration is a user-defined data type consisting of integral constants, and each integral constant is given a name. Keyword enum is used to define an enumerated data type.

```
enum typename{ value1, value2,..., valueN };
```

The typename is the name of enumerated data type or tag, value1, value2, … ,valueN are values of type typename. By default, value1 will be equal to 0, value2 will be 1, and so on. The default values can be changed using special definition numbers, as shown below:

Example 2.3.2: Examples of enumeration in C.

```
// Definition Changing the default value of enum elements

enum cardsuit{

 clubs = 0,
 diamonds = 10,
 hearts = 20,
 spades = 3

};
```

The above code defines the enumeration but does not create any instances of the variable.

A specific variable of type enum can be created, as shown below:

```
enum boolean{
 false,
 true
};
enum boolean statuscheck;
```

In the above declaration, a variable statuscheck is declared which is of type enum boolean. The variable false above has a value of 0, and true has a value of 1.

Another example is shown below:

```
enum months{
 Jan = 1,
 Feb,
 Mar,
 Apr,
...
 Dec
};
enum months months2016;
```

This example shows how convenient it is to assign values to months of the year using the construct. Since the first month was initialized to 1, the following months are increments.

2.4 Functions and Prototypes

A function in C is a set of statements that performs a particular task. All C statements must belong to a function. Functions perform the same operation each time they are called. The first function executed by a C program is the *main( )* function. The *main( )* function then executes other functions.

This is a classic example of *top-down design*, or *modular* design. Functions in C can use other functions that may have been developed by other developers or programmers. Several functions may be available to the programmer to help with typical programming tasks. For example, the C standard library contains functions for I/O, basic math, etc. Other software libraries exist for doing linear algebra, solving differential equations, statistical analysis, etc. However, in real time systems development, the developer cannot always count on all these amenities being available due to hardware and input/output device limitations. Figure 2.3 shows an example function in C with the function call code.

As per proper programming procedure, each function should have a prototype declaration. The function prototype must appear before the actual function. Prototypes inform the compiler about the types of variables (function parameters) passed to the function and the type of variable returned. If a function does not return a value, it should be declared of type *void*. Function prototypes may be included in a header file for grouping convenience. This also allows other parts of the program that

```
mainfile.c
#include <headerfile.h>

int x, y;                    /* global variables*/

void main (void) {           /* main function*/
float m = 50.0;
double n = 100.0;

function_1( CONSTANT1,m); /* call to function 1 */
y = function_2(n);           /* call to function 2 */
}
```

```
headerfile.h
/* function prototype 1 */
void function_1(int a, float b);

/* function prototype 2 */
int function_2(double c);

/* constant definition */
#define CONSTANT1        10
```

```
myfunctions.c
#include <headerfile.h>

void function_1(int a, float b){   /* function 1 */
x = a*b;
}
int function_2(double c);{      /* function 2 */
return((int) c + x);
}
```

Figure 2.3 Function Calls in the C Language with Prototypes.

may exist in different source code files to call the functions without causing a compile-time error. The figure shows a project with two source code files and one header file. The header file is included in both the source code files so that they may be compiled separately into object files. The header file in this example defines two functions and a constant. The linker will eventually combine the two object files into a single executable to be deployed on the target.

The main file declares global variables that are visible to the entire program and have to be declared outside any function. Automatic or temporary variables are defined inside a function and are not visible to other functions unless their storage location or address is passed to those functions. A careful study of the example in the figure reveals how functions accepting different types of parameters and returning different types of values can be invoked. This is the reason a function prototype is often known as an *interface*. It is left as an exercise for the reader to determine the value of variable y.

2.5 Preprocessor Directives

The C preprocessor is a part of the compiler program which performs preliminary operations before the compiler starts processing the code files. These operations may be conditionally compiling code, including content from other files, defining symbols, or something else based on the particular directive. The preprocessor responds to lines of code that start with the # symbol followed by a command known as a *preprocessor directive*. Some examples of preprocessor directives are:

```
#include <stdio.h>
#define PI 3.14159
```

Note that the #include statement doesn't end in ";" (i.e., it is not an executable statement [it is a *directive*]). The above #include statement inserts the contents of "stdio.h" into the program at the location of the directive. This is done by the C preprocessor before the program is compiled. Similarly, the #define statement defines the constant, and every instance of the string PI will be replaced by the corresponding constant value throughout the C program files where the definition holds true. An excellent use of the #define directive is to include or not include lines of code inside a file for compilation if a particular symbol is defined. This can be achieved using the #ifdef and #ifndef directives.

Both of these directives have to be closed by a corresponding #endif directive. Such constructs are extremely useful to write code that may be targeted for multiple processors or platforms or even multiple languages. Some compilers allow the passing of symbols as a command-line option, which can allow effective compilation of different lines of code within the same file. For example, whether the language C or C++ is being used can be signaled to the compiler by defining the symbol __cplusplus within the code or passing it as a command line argument to the compiler.

Table 2.2 provides a summary of the most common preprocessor directives in the C language. Another common use of the conditional preprocessor directives is *include guarding*. Header files are typically used to include common library header files; function prototypes; and variable, constant, and structure definitions used in the corresponding C source code file. It is common for popular header files to be included in multiple other header files. It is entirely possible, especially in large projects with multiple files, that common header files are included at the start of each source code file. This can result in multiple definitions of the same items causing compilation issues. Fortunately, the preprocessor can easily solve this problem by including the following code bracket.

```
#ifndef _MYFILE_H_
#define _MYFILE_H_

/* include file code with common library header files, function
     prototypes, define variables, constants and structures */

#endif // #ifndef _MYFILE_H_
```

Table 2.2 C Preprocessor Directives.

DIRECTIVE	UTILITY
#define	Definition of a symbol or macro
#undef	Remove symbol definition
#include	File inclusion
#if, #elif, #else, #endif, #ifdef, #ifndef	Conditional statements
#, ##	String formation macros
#pragma, #error, #line	Special instructions to the compiler
__FILE__, __LINE__, __DATE__	ANSI C definitions returning filename, line number, date

As can be seen, the above code, when included in a header file, will prevent it from being included more than once during the creation of a single object by the compiler.

One of the more versatile applications of the preprocessor is the definition of *macros*, or macro functions. An example of a simple macro is shown below.

```
#define DECREMENT(x) x--
```

The common template for a macro function is as follows:

```
#define MYMACRO(argument1, argument2, ...) [code ... ]
```

Multiline macros can be written using the "\" symbol at the end of each macro line indicating continuation to the preprocessor. Macros have the advantage of preventing function call overhead since the macro is inserted in a straight-line fashion into the source code. In practice, however, macros can result in a number of inadvertent errors due to this type of text insertion into the code and must be used very carefully. For example, a call such as DECREMENT(a*b) will result in an error.

2.6 Operators

Each programming language has a set of defined operators to manipulate data and addresses and evaluate expressions. The C programming language supports a number of operators, including the following common mathematical operators (the list is not exhaustive):

- addition, +
- subtraction (or multiplication by −1), −
- multiplication, *
- division, /
- modulo, % (remainder)
- increment, ++
- decrement, −−

The increment and decrement operators can be used as shortcuts to increment or decrement values by one. For example, both the statements below are equivalent:

```
k = k + 1;
k++;

The operators can be used for post-incrementing,
k = 6;
i = k++;       /*results in i=6, k=7 */
```

or pre-incrementing,

```
k = 6;
i = ++k; /*results in i=7, k=7 */
```

The language also supports the following relational operators:

- a == b, is true if a and b are equal, false otherwise
- a != b, is true if a and b are not equal, false otherwise
- a > b, is true if a is greater than b, false otherwise
- a < b, is true if a is less than b, false otherwise
- a >= b, is true if a is greater than or equal to b, false otherwise
- a <= b, is true if a is less than or equal to b, false otherwise

Table 2.3 shows a complete list of operators in C with precedence and associativity. Operator precedence rules and order of evaluation for operators with equal precedence known as associativity (left-to-right or right-to-left) can be very important. These rules are pretty standard and can also be found in any reference on the language or in the compiler documentation.

2.7 Variable Scope

The scope of a variable is the set of C statements for which the variable is defined. Variables defined within a function are *local* in scope, or *internal* variables. Variables defined outside all functions are *global* in scope, or *external* variables. The scope of a variable also refers to the visibility and extent (life) of a variable during the operation of a program. *Static* variables are defined once, and permanent storage is created for them. They exist for the entire duration that the program is running. *Automatic* variables are local and exist only within a module and are created and destroyed as needed. This is the default for all variables within a function unless they are declared to be static in nature in the program code with the *static* qualifier.

For real time systems, many variables contain values that may change frequently based on readings from a peripheral or an input/output port. A special qualifier, *volatile*, is typically used for such variables to ensure that the compiler does not optimize them since such optimization would effectively prevent updates to the variable value. The real time system will always read the current value of a *volatile* object from its source at the point it is requested rather than from a temporary register location. This is enforced even if the immediately previous instruction accessed the variable or object.

Another qualifier, *const*, prevents future modification of a variable value within a program, and any attempts to do so will be rejected by the compiler. The keywords *const* and *volatile* can be applied to structures, unions, enumerated types, or *typedef* declarations in the program. This is referred to as qualifying the declaration, and consequently, *const* and *volatile* are called type qualifiers rather than type specifiers.

Example 2.7.1: Volatile variable definitions.

Yiu provides an interesting method to declare volatile variables easily for real time systems by definitions in a header file. The definitions are shown below, and aside from "__IO," the "__I" and "__O" provide special read-only or write-only definitions that can be useful.

```
#define __I volatile /*!<˥read only˥ permissions */
#define __O volatile /*!<˥write only˥ permissions */
#define __IO volatile /*!<˥read / write˥ permissions */
```

Table 2.3 C Operators and Precedence Rules.

PRIORITY	OPERATOR	DESCRIPTION	ASSOCIATIVITY
1	()	Parentheses: grouping or function call	left-to-right
	[ ]	Brackets (array subscript)	
	.	Member selection via object name	
	->	Member selection via pointer	
	++ --	Postfix increment/decrement	
2	++ --	Prefix increment/decrement	right-to-left
	+ -	Unary plus/minus	
	! ~	Logical negation/bitwise complement	
	(*type*)	Cast (convert value to temporary value of *type*)	
	*	Dereference	
	&	Address (of operand)	
	sizeof	Determine size in bytes on this implementation	
3	* / %	Multiplication/division/modulus	left-to-right
4	+ -	Addition/subtraction	left-to-right
5	<< >>	Bitwise shift left, Bitwise shift right	left-to-right
6	< <=	Relational less than/less than or equal to	left-to-right
	> >=	Relational greater than/greater than or equal to	
7	== !=	Relational is equal to/is not equal to	left-to-right
8	&	Bitwise AND	left-to-right
9	^	Bitwise exclusive OR	left-to-right
10	|	Bitwise inclusive OR	left-to-right
11	&&	Logical AND	left-to-right
12	| |	Logical OR	left-to-right
13	? :	Ternary conditional	right-to-left
14	=	Assignment	right-to-left
	+= -=	Addition/subtraction assignment	
	*= /=	Multiplication/division assignment	
	%= &=	Modulus/bitwise AND assignment	
	^= |=	Bitwise exclusive/inclusive OR assignment	
	<<= >>=	Bitwise shift left/right assignment	
15	,	Comma (separate expressions)	left-to-right

The header file has to be included, and the preprocessor will replace "_I," "_O," or "_IO" with *volatile*. The programmer should use these types for all volatile data items such as peripheral registers which can be read or written to by software.

2.8 Program Structure

A typical C program consists of preprocessor directives followed by function prototypes and variable declarations followed by the functions or subroutines. One and only one of the functions is named *main()*, and the program always starts executing at the first line of this function. Each function contains variable declarations followed by statements. An example C program is illustrated below.

Example 2.8.1: A C program illustration.

```
#include <stdio.h>
double avg_int(int, int, int); /*function prototype*/
void main(void )
{
        double ans;
        int i, k, j;
        i = 4;
        j = 2;
        k = 5;
        ans = avg_int(i, j, k);
        printf("The answer is: %5.3g\n", ans);
}
double avg_int(int i, int j, int k)
/* this function finds the mean of 3 integers */
{
        double answer;
        answer = (float) (i + j + k) / 3;
        return(answer);
}
```

While the entire code above may be housed by the developer in a single file such as "average.c," typically, real projects are much larger and can have thousands of lines of code with many functions. In this case, there may be multiple files that house the code, each with one or more functions. It is also common practice to include the global variables and definitions in header files with the ".h" extension. So the above code could be in three files—"average.h," "main.c," and "average.c." Most IDEs allow the creation of projects that contain multiple files, including code developed by the designer, standard header files, assembly language files, and board support files for the target hardware.

2.9 C Statements

Other than variable declarations and preprocessor directives, the basic statements in the language perform value assignment to variables, arithmetic and logical operations on variables and constants, or both. Each statement ends with a semicolon. A group of statements can form a *block* that is intended to be executed sequentially. Some examples of basic C statements are illustrated in the example below.

Example 2.9.1: C Statements.

```
int i, j;              /* declaration of integers  */

i=10;                  /* assign value 10 to i          */
j=20;                  /* assign value 20 to j          */
j = (i*j)/2;           /* j equals i times j divided by 2 */
```

If only the basic statements are used, any program would be sequential without decision and control structures. C has a number of statements to facilitate these tasks. Some of these are as follows:

if Statements

These statements check a logical expression and execute the subsequent statement block only if it is true.

```
if (expression)
statement_block;
if (expression)
{
statement_block _1;
statement_block _2;
.
.
.
statement_block _n;
}
```

if ... else Statements

These statements check a logical expression and execute the first statement block if it is true and the second block if the expression evaluates to be false.

```
if (expression)
     statement_block _1;
else
     statement_block _2;
```

if ... else if Statements

These statements check a logical expression and execute the first statement block if it is true and one of the subsequent blocks based on which expression is true. If all the expressions are false, the last statement followed by the else is executed. Such statements allow for complex decision schemes.

```
if (expression_1)
      statement_block _1;
else if (expression_2)
      statement_block _2;
else if (expression_3)
      statement_block _3;
.
.
.
else if (expression_n)
      statement_block _n;
else
      statement_block _o;
```

while Loops

These statements check a logical expression and execute the subsequent statement block until it is true. The loop is exited as soon as the expression is evaluated to be false with the checks occurring before entering the block each time.

```
while (expression)
         statement_block;
while (expression {
      statement_block _1;
      statement_block _2;
      .
      .
      .
      statement_block _n;
}
```

for Loops

These statements are a form of an iterative loop that check a logical expression and execute the subsequent statement block only if it is true, with the first part of the statement initializing a condition or variable at the start and the second statement within the for loop definition, updating the condition at the end of each iteration.

```
for (statement_1; expression; statement_2)
      statement_block _3;
```

This is equivalent to:

```
statement_1
while(expression){
        statement_block_2;
        statement_3;
}
```

The most common usage is:

```
for (n = 0; n < N; n++)
            statement_block;
```

Each statement can be replaced by a block of statements, with the block being delineated by parentheses { ... statement block ... }.

2.10 Pointers and Arrays

A pointer is a variable type that can be considered to be the address of the variable. The address is the actual physical location of the variable in memory.

Example 2.10.1: Pointers.

```
int   *mu;      /*defines mu to be a pointer to an int */
int j;          /*j is an int */
mu = &j;        /* mu is assigned the address of j */
*mu = 6;        /* the integer pointed to by mu is assigned
                        a value of 6 */
```

So, & and * are inverse operators:

```
a = * (&b);   /* both are identical statements */
a = b;
```

Figure 2.4 shows a memory location with a pointer variable that contains the address or reference to that memory location. Pointers are extremely useful on a number of occasions in C and are one of the most powerful features of the language. Since they refer directly to a memory location, they must be used very carefully. Special rules may apply for incrementing and decrementing pointer variables.

Pointers make it possible to modify arguments within the calling function from the called function. Pointers are also useful to pass arrays to functions.

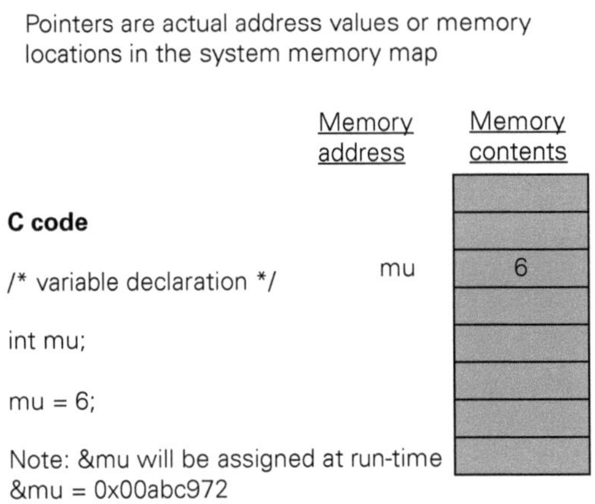

Figure 2.4 Typical Pointers as Addresses.

Example 2.10.2: Passing variables to functions by reference.

Consider the function prototype:

```
void swap(int *, int *);
```

The actual function is shown below:

```
void swap(int *a, int *b);
{
        int tmp;
        tmp = *b;
        *b = *a;
        *a = tmp;
}
```

Calling swap is achieved as follows and results in the variables being modified in the original calling function:

```
 int a, b;
swap(&a, &b);
```

Arrays are another extremely useful data structure available in C. Declaring a 1-D array is simple:

```
double x[3];
```

The array values can be initialized within a declaration, as shown below:

```
double x[] = {-9.4, 5.6, 0.4};
```

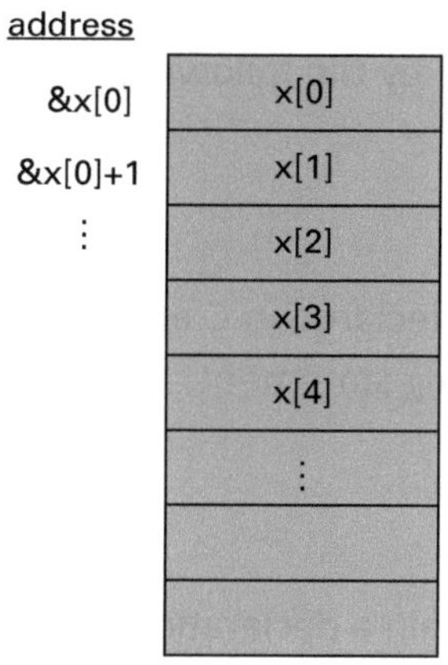

Figure 2.5 Storage of 1-D Arrays in Memory.

"x" or "&x[0]" can be regarded as pointers (addresses) to the zero element of the array. These pointers can be passed to a function.

The following example demonstrates how to pass an array to a function.

Example 2.10.3: Passing arrays to functions.

```
double avg(double[], int);
double avg(double x[], int n)
{
int i;
double sum = 0;
for (i=0; i<n; i++)
sum = sum + x[i];
return( sum/n );
}
```

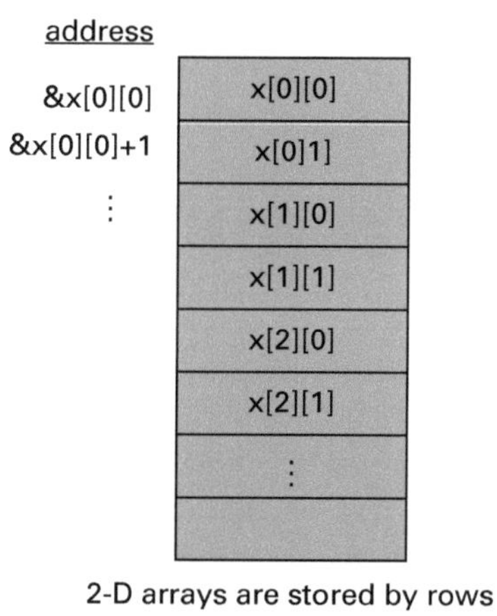

Figure 2.6 Storage of 2-D Arrays in Memory.

Invoking function avg can be achieved by the following statement:

```
mean = avg(&x[0], 5);
```

Multidimensional arrays can also be declared in C and are useful in many applications. Declaring a 2-D array is demonstrated by the following statement:

```
double x[3][2];
```

The array values can be initialized within a declaration, as shown below:

```
double x[3][2] = {{-9.4, 5.6}, {0.4, 1}, {3, 2}};
```

"x" or "&x[0][0]" can be regarded as pointers to the zero element of the array, and these pointers can be passed to a function.

The following example shows how a 2-D array can be passed to a function.

Example 2.10.4: Passing 2D arrays to functions.

```
void col_mean(double *, double *, int, int)    /* function
      prototype */
/* this function computes the mean of the columns of a matrix */
#define A(I,J) (*(a + I*n_col + J)) /* macro */

void col_mean(double *a, /*pointer to first element of matrix */
          double c[], /* result of column averages */
          int n_row, /* number of rows in A */
          int n_col) /* number of columns in A */
{
double sum;
int i,j;
for(i=0; i<n_col; i++){
sum = 0;
for(j=0; j<n_row; j++)
sum = sum + A(j,i);
c[i] = sum/n_row;
}
}
#undef A
```

Invoking the function is performed by the following statement:

```
col_mean(&A[0][0], &y[0], 4, 3);
```

This chapter provided a brief introduction to the C programming language, the intent being to cover only the major topics without going into details that can be found in any reference text dedicated to the subject.

Image Credits

- Fig. 2.1b: Copyright © Depositphotos / scanrail.

CH 3

DATA STRUCTURES

Any functional system has many different types of data associated with its constituent elements. The aggregation of different types of information into building blocks or modules makes the storage, analysis, and processing of system data very convenient. These building blocks are known as *data structures*, and many supporting algorithms exist for processing the data in these structures. The formation of data structures is critical for most computer systems and the same holds true for real time systems. In fact, most modern programming languages provide support mechanisms for data structures, or objects, as they are referred to in modern languages. The C programming language is a *structured* programming language as opposed to an object-oriented language like C++ or Java, but it has several powerful underlying mechanisms and support to implement structures. This chapter covers the concept of *structures* using the C programming language as it is commonly used and acts as a natural base leading to objects and object-oriented programming. Implementation examples with a real time operating system are also provided where appropriate.

3.1 Structures

A structure is a collection of one or more variables with its own name, typically of different types, grouped together under a single entity for convenient handling. This collection is known as a struct in C and C++. In the C++ and Java programming languages, a struct is a class with no methods. It basically defines a new kind of data type that the compiler regards as a unit. For example, the following is a new structure definition.

Example 3.1.1: Structures.

```
struct car {

     char model[72];          // model of the car
     int year;                // year car was produced
     float enginehp;          // horsepower of the engine
     int doors;               // # of car doors
     float gasmileage;        //rotational speed of motor
     };                       //struct car
```

A member of a struct is analogous to a field of a class in Java or C++. As seen, all pertinent information about a car, such as model and year, can be found in this structure and is stored in this "container." A structure can also contain other structures, making it a very powerful abstraction.

A major advantage of structures is that they can be modified to add or delete members in one single location in the code base as the product or software evolves over time. Another major advantage, as will be seen soon, is that structures can be conveniently used to pass large amounts of information to and from functions. One only has to go back to the previous chapter to note that passing multiple variables to and from functions can be quite cumbersome as C allows only one return variable. A common practice is to pass a structure or a pointer to a structure to and from a function, making it extremely convenient to pass large amounts of data during program operation.

The above definition does not declare any instances of the structure struct car. Instances can be declared by declaring struct variables, as seen below.

```
struct car x, y, z;
```

The above statement declares and sets aside storage for three variables—x, y, and z—each of type struct car. An alternative is to declare an instance in the original definition as follows:

```
struct car {

char model[72];            // model of the car
int year;                  // year car was produced
float enginehp;     // horsepower of the engine
int doors;          // # of car doors
float gasmileage;   //rotational speed of motor

} x_only;           //struct car
```

As seen, the single instance x_only is declared at the end of the definition. This type of instantiation is typically used only if a single instance is declared.

Arrays of structures can be declared similarly to declarations of variables. The declaration below is for a 25-element array of struct car.

```
struct car C[25];
```

It shows the allocation of 25 units of storage, each one big enough to hold the data of one car. Pointers to structures can be declared. As discussed earlier, pointers are just addresses and do not allocate storage for the structure (and the members) but only enough storage to hold one address.

```
struct car *c;
```

The line above declares a pointer to an object of type struct car holding one address. The pointer address variable is assigned a value inside the program, typically, when storage is assigned to it or it is used to point to an existing structure in memory.

A typical example of a structure in a real time system is a task or a thread structure. The following definition is from the ARM-RTX real time operating system available in the public domain.

Example 3.1.2: Thread or task structure in a RTOS.

/// Thread Definition structure contains startup information of a thread.

/// \note CAN BE CHANGED: os_thread_def is implementation specific in every CMSIS-RTOS.

```
typedef struct os_thread_def {

os_pthread pthread;       ///< start address of thread function
osPriority tpriority;     ///< initial thread priority
uint32_t instances;       ///< maximum number of instances of
                          /// that thread function
 uint32_t stacksize;      ///< stack size requirements in
                          ///  bytes;
                          /// 0 is default stack size

} osThreadDef_t;
```

The member types may look strange, but they are new variable types declared earlier, the procedure for which is covered in the following section.

3.2 Structure Operations

Once a structure has been declared, the program needs to access and perform operations on its members. The following example demonstrates the various ways members of structures can be accessed either from the structure variable or a pointer to the structure. Once members are accessed, they act like normal variables, and the same C language rules that apply to variables apply to the structure members. Let p and q be instances of the motor structure declared in the previous section.

```
struct car p;
struct car q[10];
```

The various members can then be accessed as follows:

```
p.year          - is the year
p.enginehp      - is the engine horsepower
p.doors         - is the number of doors
p.gasmileage    - is the gas mileage
```

Also, for the structure array variable,

```
q[i].year       - is the year of the ith car
q[i].enginehp   - is the horsepower of the ith car
```

In case, a pointer to the structure is required

```
struct car *p;
```

The structure members can be accessed as follows:

```
(*p).year       - is the year of the car pointed to by p
(*p).enginehp   - is the horsepower of the car pointed to by p
```

The parentheses are used to dereference the address first because the "." operator has a higher precedence than the unary operator "*." For example, in expressions such as the one below

```
m.enginehp * m.gasmileage
```

the "*" operator acts as the multiplication operator and is used to multiply the two members. If the parentheses are not used, the compiler will try to access p.year first (is this possible?) and then dereference the result.

The (*p).member notation is cumbersome and requires too many keystrokes. The construct is so widely used that a special notation was invented that is ubiquitous in C and C++:

```
p->member, where p is a pointer to the structure.
```

Using this operator provides a much better interface to the programmer:

```
struct car *p;
p -> year       - is the year of the car pointed to by p
p -> enginehp   - is the horsepower of the car pointed to by p
```

The following are some other common operations on struct variables.

Example 3.2.1: Structure operations.

```
struct car p, q, *s;
struct car m = {"Ford", 2008, 200.0, 2, 21.5}; /// initialization
p = q;  /// Copy/assign one structure variable to another
s = &p;                        /// Get address
p.year;                        /// Access members
s -> enginehp;                 /// Access members
void mysub(struct car m)       /// pass structure to a subroutine
void mysub(struct car *m)      /// pass structure pointer to a
                                         /// subroutine
struct car mysub(void) /// return structure from a subroutine
struct car *mysub(void) /// return structure pointer from a
                                    /// subroutine
```

It is important to remember that passing an argument by value is an instance of copying or assignment and that passing a return value from a function to the caller is an instance of copying or assignment.

```
struct car mysub(struct car g) {
struct car h = g;
     ...;
     return h;
}
```

In the above example, the structure passed to the function mysub will be copied into the local variable struct car g, after which it will be copied into the structure that is equated to the function return. The following call will copy the structure q to p (although a roundabout way of doing it).

```
p = mysub(q); /// call to the function defined above
```

There has been some discussion about whether passing a structure pointer to a function is more efficient that passing the structure variable itself. Copying is very fast on modern computers; however, creating an object with dynamic memory allocation and assigning a pointer is not as fast if it is desired to have the object passed or returned by value.

It is important to note certain rules about structures and pointers. A declaration of a structure without details of the internal members is allowed. For example, the following declaration is legal and can be found in many implementations:

```
/* in a .c or .h file */
struct item;
struct item *p, *q;
```

This is called an *opaque type* because the program can use pointers to items but cannot see into items. The program cannot define any items and is not allowed to dynamically allocate storage for the structure. However, the structure has to be defined and implemented in some other file as follows:

```
...     /* In another file */
struct item {
        int member1;
        float member2;
        struct item *member3;
};
```

This has many advantages. The implementer of item can change the definition without forcing users of pointers to change their code!

One more note about structure declarations is in order. The following declaration is not legal:

```
struct car {

char model[72];             // model of the car
int year;                   // year car was produced
float enginehp;     // horsepower of the engine
int doors;          // # of car doors
float gasmileage;   //rotational speed of motor

};                  //struct car

car m;              /// illegal

car *p;             /// will produce compilation error
```

It is necessary to declare any instances as follows:

```
struct car m;       ///correct declaration
struct car *p;      ///will compile fine
```

Due to this rule, the typedef declaration is often used to rename the type for convenience.

Example 3.2.2: Type definitions in C.

```
typedef struct car Car;
Car m, n;
Car *p, r[25];
Car function(const Car m; ...);
```

The typedef allows leaving out the word "struct" and, in general, can be used to rename any type, not only structures. This has many advantages other than convenience—it can be used to clarify the purpose of the type, produce clean readable code, and promote portability across platforms. Type definitions can also be used for pointers:

```
typedef char *String;
```

or for normal C variable types to define the exact size for the programmer's convenience.

```
typedef int size_t;
typedef long int32;
typedef long long int64;
```

This is very common in *C* and *C++* real time programming to create portable code as the variable sizes for short and long integers depend on the processor and may change from platform to platform. The particular details for the target processor can be defined once in a header file and can even be selected based on compiler flags or definitions (e.g., #define ARM32 or –DARM32).

3.3 Dynamic Memory Allocation

In the C programming language, memory can be managed in different ways: static, automatic, or dynamic. Each method is closely related to the scope or lifetime of the variables or structures declared in the program while the program is running. The program typically exists in main memory at run time. Static variables are allocated in main memory and persist for the lifetime of the program. Automatic variables are typically declared within functions and created when the function is called and deleted upon return. They are typically allocated storage on the stack which grows upon allocation and shrinks later. The size of allocation for static and automatic variables must be constant when the program is written since the compiler does not allow variable size for these types (except in C99, which allows variable-length automatic arrays). This can be a problem if the required size is not known until run time, which can happen if data of arbitrary size is being read from the user, process or file. In this case, fixed-size data objects are inadequate as run time errors may occur if sufficient storage is not allocated.

The lifetime of allocated memory is also important as neither static nor automatic variables are adequate for all situations. Automatic variables do not persist across multiple function calls while static variables persist for the life of the program. This may not be ideal for larger structures and for real time systems with limited memory. The programmer may require greater flexibility to allocate memory to variables at run time.

Dynamic memory allocation is a practical solution to avoid such limitations and allows explicit and flexible memory management. It is typically achieved by allocating memory from the free space, known as the *heap*. Most systems have an area of memory structured for this purpose. The library function malloc() is used to allocate a block of memory on the heap in C. The program can then access this block of memory via a pointer that is returned by the function. When the allocated memory is no longer needed, the pointer is passed to the free() function which deallocates the memory to be used for other purposes.

Example 3.3.1: C dynamic memory allocation functions.

The C dynamic memory allocation functions are defined in the "stdlib.h" header (cstdlib.h in C++). The following are the functions typically available for dynamic memory allocation:

```
Function       Description
==================================================================
malloc()       allocates the number of bytes specified

realloc()      increases or decreases the size of the block of
               memory specified reallocating as needed

calloc()       allocates the number of bytes specified and
               initializes them to zero

free()         releases the block of memory specified back to the
               system
```

The two differences between the malloc() and calloc() functions are: the former takes a single argument which is the amount of memory to be allocated in bytes while the latter needs two arguments—the number of variables to be allocated and the size of a single variable in bytes. The former does not initialize the allocated memory while the latter guarantees that all allocated bytes of the memory have been initialized to zero.

Example 3.3.2: C dynamic memory allocation usage.

The following examples demonstrate usage of these functions. The creation of an automatic array of ten integers in C is a simple declaration. Once it is declared during compilation, the size cannot be changed:

```
int array[20];
```

In order to create an array such that its size can be changed at run time, or, in other words, dynamically, the memory allocation functions must be used:

```
int *array = malloc(20*sizeof(int));
```

malloc() needs as an argument the number of memory bytes required, and the sizeof(int) function is extremely useful to determine array element size even for arrays of complex structures. malloc() will request the total number of bytes required from the system and then will return a pointer to the memory block returned by the system. It is entirely possible that the malloc() call will fail, in which case, the system is not able to service the allocation request, and the result is the return of a null pointer. The standard programming practice is to check for such an eventuality as follows:

```
int *array;
array = malloc(20*sizeof(int));
if (array==NULL) { /* check for null pointer return */
 fprintf(stderr, "malloc call failed\n");
 return(-1);
}
```

Once the dynamic array is not needed any longer, the following call returns the memory to the free store (typically heap memory):

```
free(array);
```

The memory block returned by malloc() is not initialized to zero and can contain random numbers or the remnants of previous computational data. If initialized variables are required or important for the program, the command calloc() will conveniently allocate and clear the memory in a single step:

```
int *array = calloc(20, sizeof(int));
```

The above line of code requests a block of memory large enough to hold 20 integers and will initialize to zero all successfully allocated bytes in that memory space.

It is important to note that the sizeof() operator is extremely useful to determine the amount of memory storage that is required to dynamically allocate many types of data structures that are described in further detail later in this chapter. A sizeof(char) call will always return a 1, but this can be extended to other types of structures. The following call will allocate enough storage for an array of 10 structures, with the return value for sizeof(struct car) being the number of bytes required for one structure:

```
struct car *p;      /* pointer to car structure defined
                       earlier */

p = calloc(10, sizeof(struct car));
                    /* storage for an array of 10
                         structures */
p[0].model = "Ford";  /* access struct element */
```

One of the major advantages of this approach is not having to rewrite the code if the structure definition changes due to addition or deletion of members. Another advantage is that the compiler sometimes pads structure elements with empty bytes for memory alignment purposes, and the operator will automatically compute the actual structure size.

There are a number of reasons to be careful about using dynamic memory allocation routines, particularly in an embedded system. Some of the pitfalls may include memory fragmentation—embedded systems can run for long periods of time and make multiple calls to obtain and free memory, which may cause waste of memory due to fragmentation (similar to a hard disk). Speed can also be an issue—dynamic memory allocation is relatively slow (and gets slower as the memory gets fragmented). If the same dynamic memory is being used for different threads, then allocation and freeing routines need to perform locking, which, in turn, can cause problems servicing interrupts, quickly affecting system response time and interrupt latency.

Dynamic memory allocation makes it very difficult to debug, particularly with some of the limited/primitive debug tools available on an embedded system. The modern tools currently available have improved this situation considerably. One of the most important things to remember for any system is to ensure that allocated memory is freed once it is no longer needed. This involves matching or pairing allocation and free calls. If memory is not freed correctly, it causes the program footprint to keep growing—a process known as a *memory leak*. If the memory is freed at the incorrect time, there may be *dangling pointers*—variables pointing to memory being used by another variable or structure. In either case, the system will display erratic behavior such as a crash.

It is important to judge whether dynamic memory allocation is really required for a real time system. Dynamic memory allocation allows the system developer to reuse the same memory to do different things at different times. Simpler embedded systems typically tend to do the same thing all the time (except at startup). However, in today's complex systems with multiple threads and tasks, dynamic memory allocation may be difficult to avoid.

3.4 Unions

A *union* is like a structure but a totally different construct. It signifies a single physical storage entity or bin that may hold different types of variables at different times. Members are overlaid on top of each other within the same physical space. At any given time, only one of its members is stored, and the storage is enough to hold the largest member. The following example will clarify the concept.

Example 3.4.1: C union.

```
union {
  int ivalue;
  float fvalue;
  char *svalue;
} u;
```

The above union has only four bytes allocated to it, assuming that integers, floats, and address pointer are all of the same length on the target system (4 bytes). Different types of variables can be stored in those four bytes, i.e., different logical entities in the same physical space. It is the programmer's responsibility to keep track of the type stored in a union at any given time during program operation!

A union can provide a powerful construct for many applications, as seen in the example below:

```
struct taggedItem {
enum {iType, fType, cType} tag;
union {
      int ivalue;
      float fvalue;
      char *svalue;
}u;
};
```

The members of the structure above are the enumeration and the union.

```
enum tag;
union u;
```

The value of the tag can be used to determine which member of the union to use.

Unions are used infrequently as compared to structures—mostly in operating systems and device drivers. Detailed implementations are also frequently found in real time systems in order to access registers defined by the hardware.

3.5 Stacks

A stack is an abstract data structure that is used to store objects with specific rules to add and remove them from the structure. Stacks are last-in-first-out (LIFO) data structures. The best way to visualize a stack is a bin that is open at the top with the further restriction that only objects that fit the opening exactly are allowed to be placed inside it. Initially, the bin is empty, and the program starts filling it with objects. However, only the last object that was placed inside the bin can be removed since it is the only one that can be accessed from the opening. If the first object placed in the bin needs to be accessed, all the objects on top have to be removed sequentially in order to be successful. This abstraction, while it may seem contrived, is extremely powerful in many computer applications and has been used for decades. Calculators that employ reverse Polish notation (operands entered first, followed by operation) use a stack structure to hold operand values. Stacks can also be used in search algorithms for *backtracking*. Many virtual machines are also stack-oriented, including the Java Virtual Machine. Several programming languages use stacks to store procedure or method data that is typically local in nature. Any space required for local data items, including function parameters, are allocated from the stack just before the procedure is entered and deallocated upon procedure exit. The C programming language is commonly implemented in this manner. Recursive procedure implementation makes heavy use of the stack.

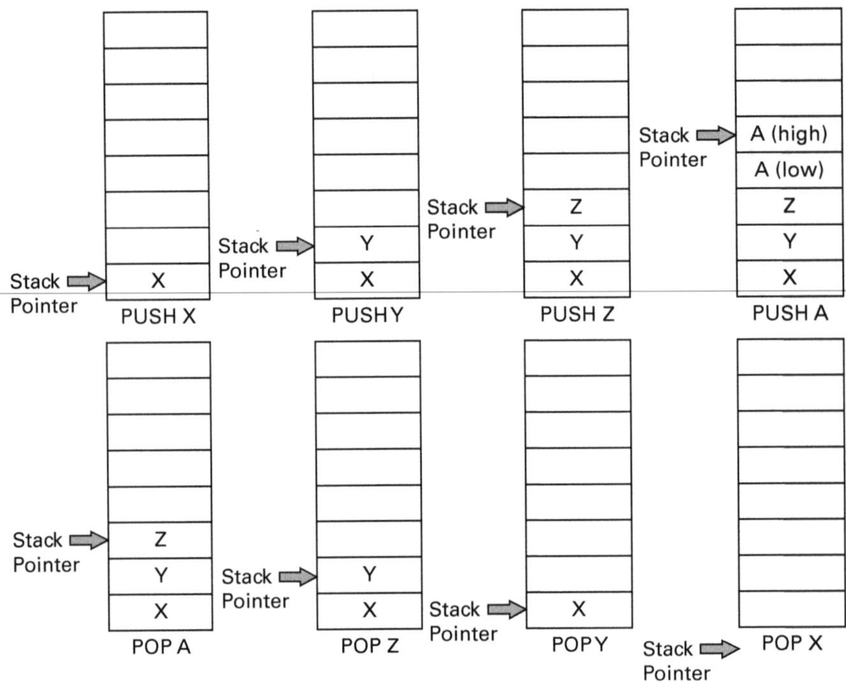

Figure 3.1 A Typical Stack Implementation.

Stacks can be implemented in hardware or software. Most target processors provide hardware support for stacks in the form of a *stack pointer* that is a register which holds the updated address of the stack. There are also native instructions that can place and remove items on the stack, known as *push* and *pop* instructions. The instructions automatically change the *stack pointer* based on the size of the items placed on or removed from the stack. The *stack pointer* has to be initialized by the programmer to point to a valid section of memory such as RAM. Stacks are ubiquitous in most processors as they are used to store the "context" of a processor or task when it is interrupted by a hardware *interrupt* signal or some other method. A stack is also the most convenient method to store the return address whenever a subroutine or function call is made in the program. It can be, thus, thought of as a temporary storage space with strict rules of operation.

A stack is usually implemented to have a bounded capacity due to physical considerations. If the stack does not contain enough space to accept another object, it is full and considered to be in danger of overflow. An underflow condition can also occur if the stack is totally empty and a pop operation is performed. Environments that rely heavily on stack operations may provide additional conveniences, for example:

- Duplicate: the item at top of stack is duplicated and placed above the current top item.
- Peek: the top stack item is inspected (or returned) without changing the stack pointer.
- Swap or exchange: the top two items on the stack are exchanged.

In theory, stacks can have unbounded or very large capacity with cheap memory availability. Typically, dynamic allocation has to be used in this case with extra storage being requested and allocated as required. Stacks can also be implemented in software using arrays. The following sample code shows a stack implementation in C.

Example 3.5.1: C Language stack implementation.

```
#define STACK_SIZE 100
struct StackC {
 int  data[STACK_SIZE];
 int  size;
};

typedef struct StackC Stack;

void Stack_Init(Stack *SS)
{
 SS->size = 0;
}

int Stack_Top(Stack *SS)
{
 if (SS->size == 0) {
  fprintf(stderr, "Error: stack is empty\n");
  return(-1);
 }
 return(SS->data[S->size-1]);
}

void Stack_Push(Stack *SS, int d)
{
 if (SS->size < STACK_SIZE)
  SS->data[SS->size++] = d;
 else
  fprintf(stderr, "Error: stack is full\n");
}

void Stack_Pop(Stack *SS)
{
 if (SS->size == 0)
  fprintf(stderr, "Error: stack is empty\n");
 else
  SS->size--;
}
```

A stack can also be implemented with a *list* structure. The top of the stack, in such a case, is a pointer to the *head* of the list, and a counter tracks the list size. The following section will introduce linked lists, upon which it will be evident that a stack can be easily implemented.

3.6 Linked Lists

While arrays of variables or structures can be utilized to implement almost all data structures, they have advantages and disadvantages associated with them. Arrays have to be declared at compile time, thereby, having a fixed size at run time. This can be good for determinism but can be a problem if array bounds are not sufficient to achieve program goals or need to be exceeded. Some operations, such as adding data to a stack, list, or queue, are much easier and faster using arrays. In these cases, typically all that need be done is add data and increase the pointer to the end of the queue, for example. Most real time systems require other operations such as deletion of an element, insertion of an element, or prioritization of elements in a list. In such cases, arrays can be difficult to work with and time-consuming from a computational perspective. Luckily, alternative solutions such as linked lists have already been formulated and tested by computer scientists and have been in operation for some time. Linked lists have several advantages:

- Dynamic data structures allocate needed memory during program operation
- Insertion and deletion operations for nodes can be easily implemented
- Linear data structures, including stacks and queues, can be easily executed
- Access time can be reduced and structures can expand in real time

It is best to explain the concept of a linked list with an example.

Example 3.6.1: Singly linked list.

The following is a definition of a structure with a pointer to another structure of the same type:

```
/* implemented in .c or .h file */
typedef struct _item Item;     /* item structure type definition */
Item *p, *q;                   /* pointers to item structures */
/* In another file or location */
struct _item {
        char *mydata;
        Item *nextItem;
};
```

The above structure is known as a *self-referential,* or a *recursive* data structure. In such a data structure, one structure item can point to another item, which can then point to another one, which can point to yet another one, and so on, forming a list of items. Each item in a linked list is called an *element,* or *node*. The field in the node that holds the address of the next node is called the *next link*, or *next pointer*. The remaining fields are popularly known as the *data*, *information*, *value*, *cargo*, or *payload* fields. The *head* of a list is the first node, and the *tail* of a list refers either to the rest of the list after the head or to the last node. If the list is not empty, the last node will have a NULL pointer as the *nextItem field.

Figure 3.2 shows an example of a singly linked list. A doubly linked list is shown in the Figure 3.3. Each node in a doubly linked list points to the next node and the previous node.

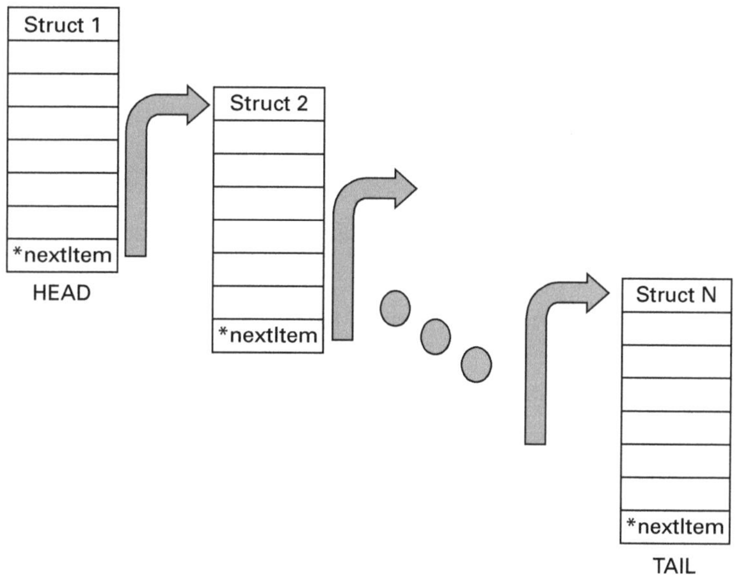

Figure 3.2 A Singly Linked List Structure.

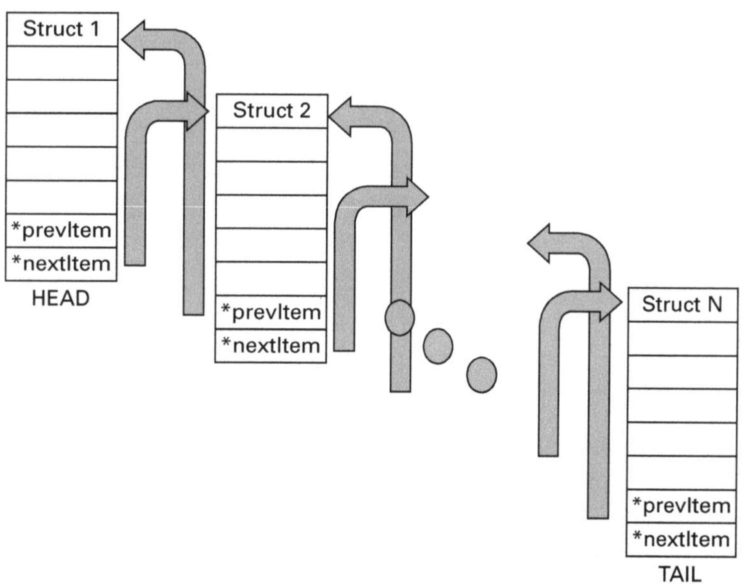

Figure 3.3 A Doubly Linked List Structure.

Example 3.6.2: Doubly linked list.

The example code for a doubly linked list structure is shown below:

```
/* implemented in .c or .h file */
typedef struct _item Item;      /* item structure type definition */
Item *p, *q;                    /* pointers to item structures */
/* In another file or location */
struct _item {
        char *mydata;
        Item *previousItem;
        Item *nextItem;
};
```

For a doubly linked list, there is an extra pointer to the previous node, as shown in the structure. For the head node, there is a NULL pointer in the previous item field, and the next item pointer for the last node is the same as that for a singly linked list.

The advantages of linked lists have been enumerated earlier in the section. New nodes can be added or deleted on the fly with dynamic memory allocation. Linked lists do have some disadvantages, and the real time engineer needs to be aware of them to make a judicious design choice vis-à-vis arrays for systems:

- They use more memory because of pointers that require extra storage space.
- They require sequential access, and nodes must be read in order from the beginning.
- Extra time may be necessary to access individual elements in the list since the nodes may not be contiguous.
- Reverse traversing can be cumbersome especially for singly linked list, and although doubly linked lists are easier to read, memory is wasted by allocating space for the extra back pointer.

Overall, however, they offer an excellent design choice for many real time system structures such as task lists, queues, mailboxes, and priority lists. For example, items can be deleted and or added without the necessity of moving to the head of the list. Figure 3.4 illustrates these actions.

3.7 Queues

A *queue* is defined as a list of data items, commands, etc., stored such that they are retrievable in a definite order, usually in the order of insertion. Just like the stack, a queue is an abstract data type with the difference in operation is that it is typically a FIFO or a first-in-first-out data structure. The principal operations allowed on the structure are adding an element, known as *enqueue,* or removing an element, known as *dequeue*. A normal queue acts like a buffer, and once an element has been added, it has to wait until all elements in front of it have been *dequeued*. Figure 3.5 shows an example of a queue.

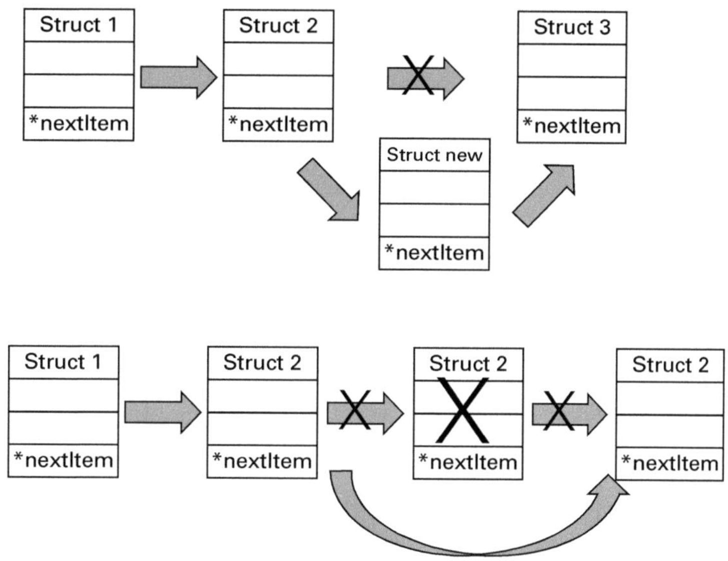

Figure 3.4 Insertion and Deletion from a Linked List.

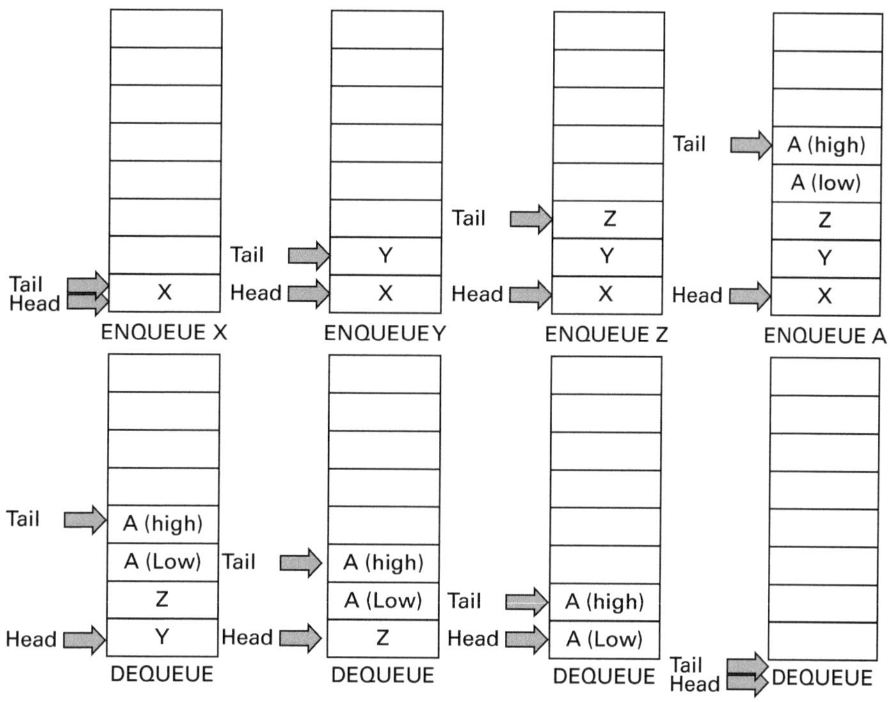

Figure 3.5 A Queue Structure.

A variation of the normal queue structure is the prioritized queue. It differs from a normal queue in that each element may have a priority level associated with it. The dequeueing process is the same as a normal queue. However, the enqueueing process may place an element in the queue based on its priority level. The figure below shows an example of a prioritized queue.

The implementation of a queue can be achieved with either an array, array of structures, or a linked list. A queue with a fixed number of items is known as a bounded queue. Implementation of queues

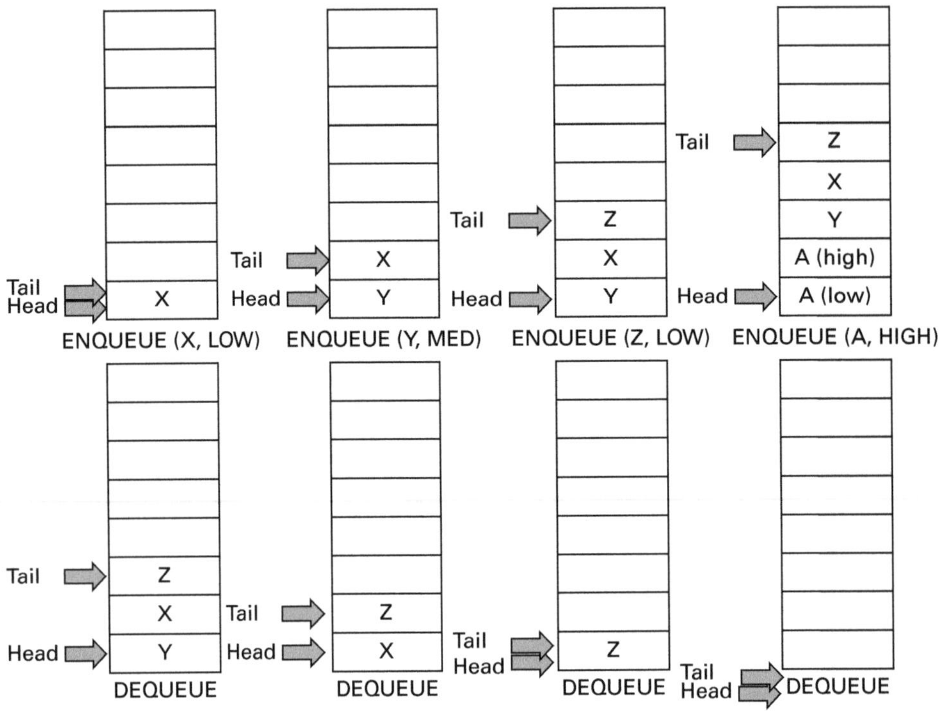

Figure 3.6 A Prioritized Queue.

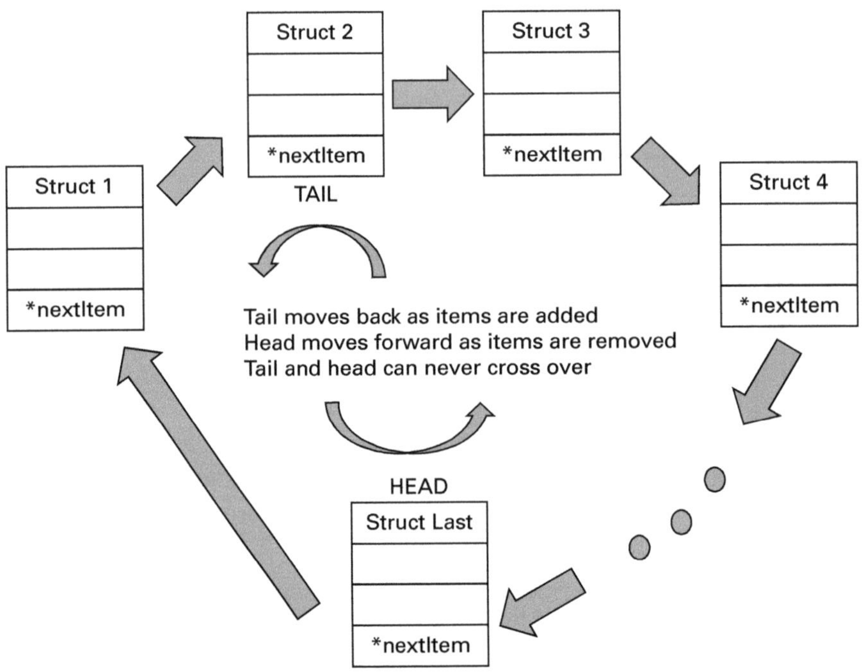

Figure 3.7 A Circular Buffer.

with an array can be cumbersome and inefficient from a computational time perspective for the queue operations. A linked list lends itself very well to the implementation of a queue since items can be added or removed dynamically. A doubly linked list is very efficient for queue implementation. A singly linked list can also be used but is less efficient unless a pointer to the first and last node is maintained.

Another method for the efficient implementation of bounded queues is the circular buffer, in which the last item is linked to the first item. There is, typically, a read pointer at the head of the queue and a write pointer at the end of the queue. A popular alternative for digital signal processing (DSP) applications is a simple dual buffer system also known as the ping-pong buffer. Typically, such a system has analog-to-digital and digital-to-analog converters that write data into a buffer for processing and read data from a buffer after processing, respectively. The use of a dual buffer at each end allows the DSP algorithm or other system components to read data from one buffer (known as the ping buffer) while the other buffer (known as the pong buffer) is being filled with data. Once the pong buffer is full of data, the system starts writing to the ping buffer, and data extraction for processing is switched to the pong buffer. Such a system improves the quality of the system as the buffers can smoothly process the data at a consistent rate without interruptions.

In real time systems queues, prioritized queues and circular buffers can be used for a number of reasons, such as requests for resources, task lists, passing messages or data between tasks, and data storage. They are especially useful for such communication from a single process to a single process, single process to multiple processes, and multiple processes to a single process. Several algorithms are available for implementing these data structures.

CH 4

REAL TIME OPERATING SYSTEMS

The design of real time systems has evolved over time with successively more sophisticated approaches culminating in a full-fledged, real time operating system. It is worthwhile to look at the various design approaches implemented by system designers to meet real time requirements. There are three general approaches to task scheduling: *ad-hoc scheduling*; *deterministic scheduling,* using a cyclic executive; and *non-deterministic, priority-driven scheduling,* using a multitasking executive. The design decisions are based on a number of factors, and this chapter looks at these approaches.

4.1 Basic Solutions

Ad-hoc scheduling is the simplest form of task scheduling if it can be called task scheduling at all. The ad-hoc approach should not be discounted; for rather straightforward processing with no specific periodic requirements, it can produce satisfactory results when used with interrupts to provide the processing associated with external events. In ad-hoc scheduling, the tasks are functional program units in which the dependencies are limited to precedence relationships. A precedence relationship is simply a requirement that one task be completed prior to the execution of another. The scheduling of the tasks is merely a matter of having a task, upon completion, start the following task. This process continues until all tasks have completed, at which time, the first task is executed and the process repeats.

Even though task execution is repetitive, it is not periodic. That is, the repetition does not occur at a precise rate but, rather, a task repeats whenever its predecessor task completes execution. Thus, there can be no specific periodic requirements placed on task execution. The program may essentially be written as an endless loop with each task executing to completion in some predetermined sequence and with asynchronous events handled by interrupt processing. The following subsections describe the various types of basic solutions to meet real time system deadlines.

4.1.1 Polled-Loop Systems

These systems represent the most basic type of architecture for a real time system that involves polling devices attached to the system continuously and servicing them upon request. If the attached device or subsystem does not need servicing, the main loop continuously keeps checking for a service request event, which can be a flag or signal in the software. Depending upon the type of input event, sometimes, a small delay is introduced between successive attempts at checking the status to achieve *debouncing,* especially if the event is slow in nature, much like a human interface device. In any case, the event flag is typically cleared upon servicing so that the system is ready for the next event or *bursts* of events.

A polled-loop system can be thought of as the simplest form of a real time kernel. It is simple to design and debug as it does not require interrupts. Polled-loop systems are used for fast reaction to single events that require a guaranteed response time. The figure below shows the design flowchart for a simple polled-loop system that interfaces with multiple peripherals. One of the major disadvantages of a polled-loop system is that the system and, thereby, the processor or CPU is dedicated to the status loop, thus, making it impossible to do other tasks or even enter a "*power-saving*" or sleep mode. Also, as in the second case above, as the system gets more complex, it is not possible to guarantee that the peripherals will be serviced in the correct order or priority level which can be a major issue.

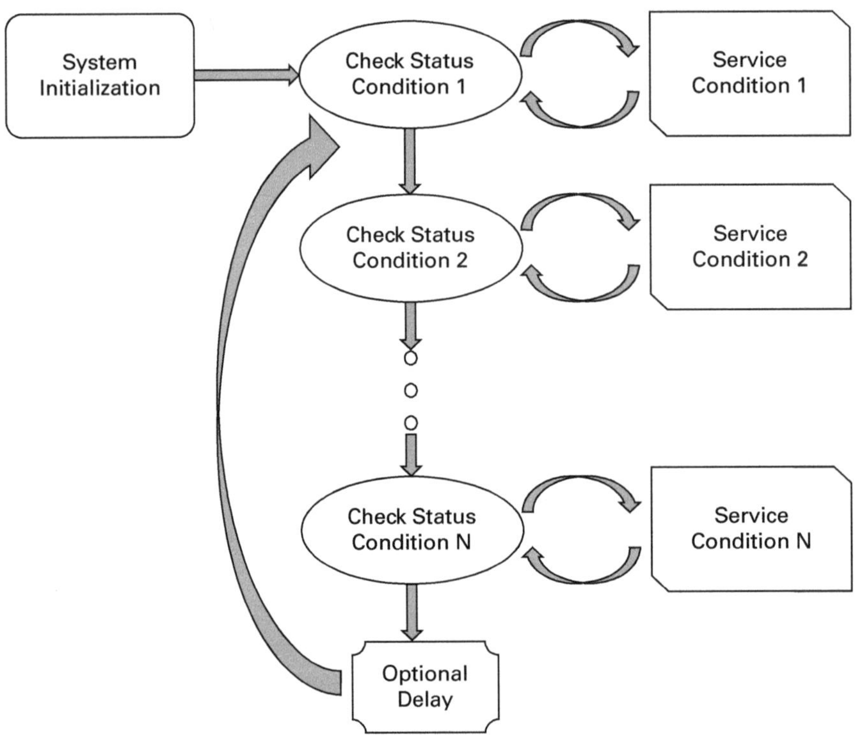

Figure 4.1 Polled-Loop Systems.

4.1.2 Interrupt-Based Systems

A more sophisticated approach to designing a real time system is to utilize the interrupt processing capability of the processor. Interrupts come in three flavors: hardware interrupts, software interrupts, and exceptions. When an interrupt occurs, the processor is forced, by design, to temporarily suspend the currently executing program and *service* the interrupt. A hardware interrupt is an electrical signal generated by a peripheral device and sent to the processor via a dedicated pin or line. A hardware interrupt is truly random or asynchronous in nature since it is not possible to predict when it will occur in the code sequence. A software interrupt is typically inserted by the programmer in the code and has the same effect as a hardware interrupt. Due to the predictable nature of a software interrupt based on its location in the code, it is considered to be non-random or synchronous in nature. An exception is an interrupt that is internal to the CPU and triggered by an attempt to perform an unexpected or illegal operation. Often, this interrupt is based on the operands and, as a result, is random or asynchronous in nature. Regardless of the type of interrupt, the processor services them via special subroutines known as ISRs or *interrupt service routines*.

Typically, each type of interrupt on a processor can have a separate routine handling it. These routines have to be placed by the system designer in program memory, and a special table known as the interrupt vector table (which is a set of pointers to the ISRs) must be populated with the location of the ISRs. Many systems also allow priority levels to be attached to interrupts, wherein an ISR for a lower priority interrupt can itself be interrupted by a higher priority one. When an interrupt occurs, the current "context" of the processor is saved on the stack, and the ISR is executed. Upon completion of the ISR, normal processing can resume once the *context* is restored. The context usually includes contents of general registers, program counter, and any other information necessary to restore execution upon return. The figure below shows the design flowchart for an interrupt-based system.

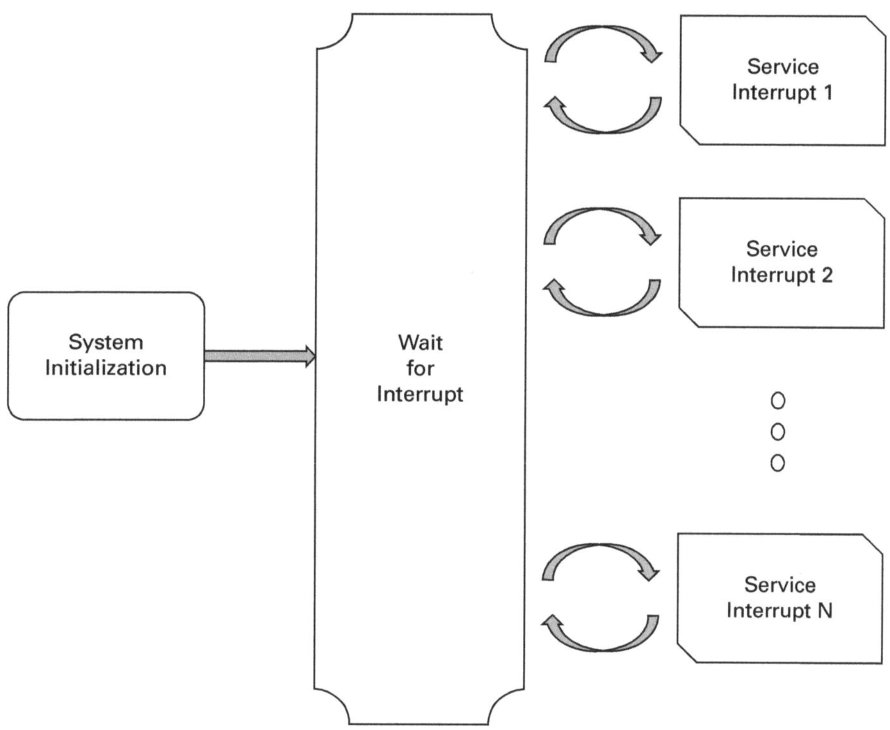

Figure 4.2 Interrupt-Based Systems.

As seen above, a pure interrupt-based system initializes and then enters sleep mode whereupon it idles until an interrupt wakes it up. While such systems are effective in nature, there are some drawbacks. All interrupts with the same or lower priority are typically disabled inside an ISR. As a result, it is not advisable to spend too much time inside an ISR since even important events like timer signals can be disabled during this time, possibly creating incremental errors. Another problem is that synchronization mechanisms cannot be utilized inside an ISR. Such systems have to be very carefully designed to make sure they do not *hang* while waiting for a response from another resource.

4.1.3 Finite-State Machines

An alternative approach to the above is to create a *Finite-State Machine* (FSM) and allow the system to transition between different states. A finite-state machine is a mathematical form of computation based upon a hypothetical machine that can store the *state* of a system at any particular time. The state of a system at time t is the totality of all variable quantities within the system at time t so that the condition of the system can be precisely reconstructed at some other time T. The system itself executes by moving from one state to the next. This movement from one state to the next is called a *state transition*. The transition itself is a function of only two things—the current state of the system and the input to the current state.

Finite-state machines, then, consist of four main elements:

- States which define behavior and may produce actions.
- State transitions which are movements from one state to another.
- Rules or conditions which must be met to allow a state transition.
- Input events that are either externally or internally generated, which may possibly trigger rules and lead to state transitions.

Finite-state machines must have an initial state, which provides a starting point, and a current state, which remembers the product of the last state transitions. Received input events act as triggers, which cause an evaluation of some kind of the rules that govern the transitions from the current state to other states. The best way to visualize a FSM is to think of it as a flow chart or as a directed graph of states. Figure 4.3 shows the directed graph of states representation of a simple system which transitions between different states represented as *parts* in the process code. The system shown transitions very systematically between the two processes, but more complex transitions to different process parts can be easily designed.

The finite-state machine is one of the most popular design patterns in embedded systems. Many applications from simple home appliances to complex communication systems implement event-based state machines. As noted above, the finite-state machine is made up of multiple states. At any point in time, the system is in one state, and an event triggers certain actions in that state along with a possible change in state. The event could be due to an interrupt in the system, an RTOS signal, a timer expiry indication, or an input or indication from another module in the system. Various system tasks can be divided into subtasks and then a state associated with each subtask. Once a subtask is completed, the program changes the state and relinquishes control to the main loop which then decides the next subtask to perform based upon a switch statement associated with the designated states. Figure 4.3 shows the flow of operation for a system based on a finite-state machine. Such a design can be used

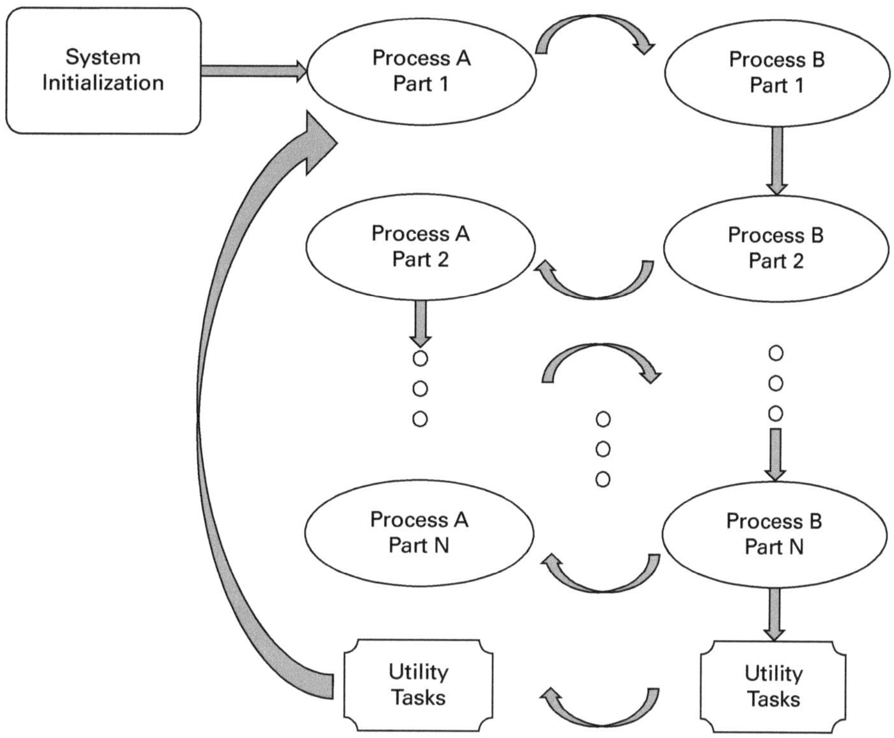

Figure 4.3 Finite-State Machines.

in conjunction with interrupts or polling to form fairly sophisticated systems. However, they can be difficult to modify and debug once they are designed.

4.1.4 Combination Systems

A number of variations of the above concepts may be found in the systems being deployed in industry. Figure 4.4 shows a system with a combination of polling and interrupts. The interrupt sets a software flag which is then handled by the polling routine. Such a design prevents a lot of time being spent inside an ISR and also lets low-priority interrupts be serviced and not *crowded* out by frequent occurrences of higher priority interrupts. Another type of system depicted in The figure below is a *hybrid* system where some of the routine tasks are processed by the main loop and known as *background* or *cleanup* tasks. These tasks are not time-critical and may include items like low-priority self-testing, display updates, logging, and interfaces to slow devices. Typically, such combination systems are the result of strict timing requirements, desire for simplicity, or even to avoid the cost and overhead of a full-fledged operating system. Such systems are typically known as *home-grown* systems and evolve over time. They can be proprietary and difficult to maintain since no common design concept has been used, and several designers may have contributed to it over time. They can be fairly effective and are still in operation. Whenever a higher degree of reliability, scalability, and maintainability is desired, a full operating system is used.

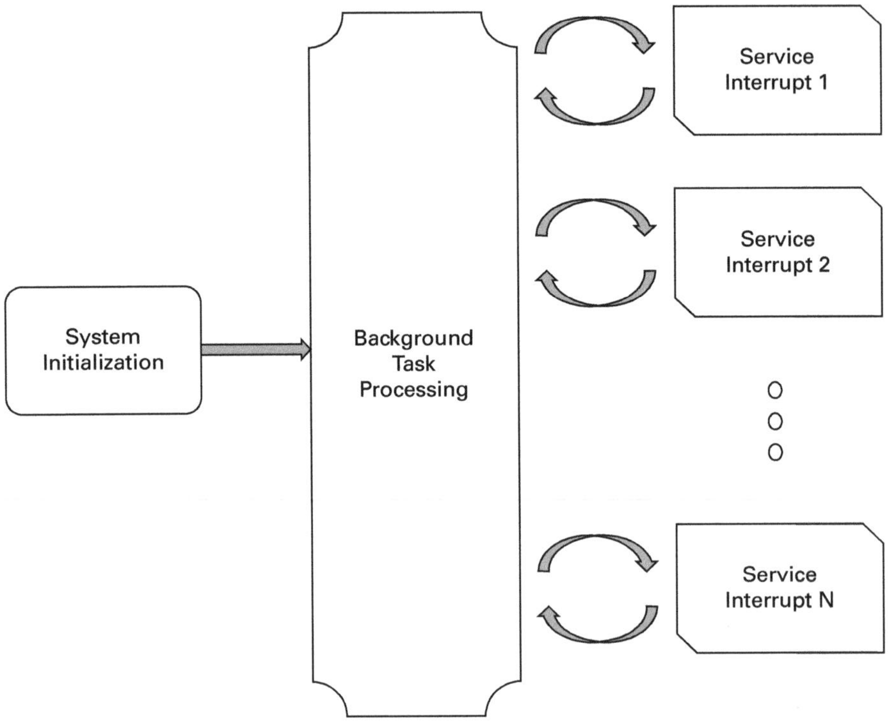

Figure 4.4 Combination Systems.

4.2 The Cyclic Executive

One of the most common approaches used when the system specifications require strict periodicity is a priority-driven scheduling solution known as the *cyclic executive*. The cyclic executive approach has the advantages of providing deterministic scheduling and low implementation overhead. This is because the cyclic executive control software is typically small and fast. The cyclic executive approach works extremely well for systems that are:

- Periodic—with a high periodic content
- Predictable—the tasks have predictable execution times
- Synchronous—have relatively few asynchronous requirements
- Deterministic—have fixed execution sequence requirements

An example of such a system is an inertial navigator, in which the inertial instruments provide data at precise periodic rates, and this data must be processed prior to the arrival of the following data. Asynchronous events, such as control panel entries by the operator, are relatively rare. The execution times of the various tasks are fairly predictable since they involve about the same amount of calculation each time they are executed.

The main goal of a cyclic executive is to meet the strict timing requirements imposed upon the software. For such a scheduling strategy, the tasks are deterministically scheduled on a periodic basis. The tasks are scheduled for execution by an interrupt-driven control program called a cyclic executive.

The cyclic executive program has a scheduler that bases its timing upon a hardware timer subsystem that generates periodic interrupts. Figure 4.5 shows a cyclic executive system with periodic tasks driven by timer interrupts. Such timer subsystems are an integral part of the embedded computer or CPU, and the interrupts correspond to *ticks of the real time clock*.

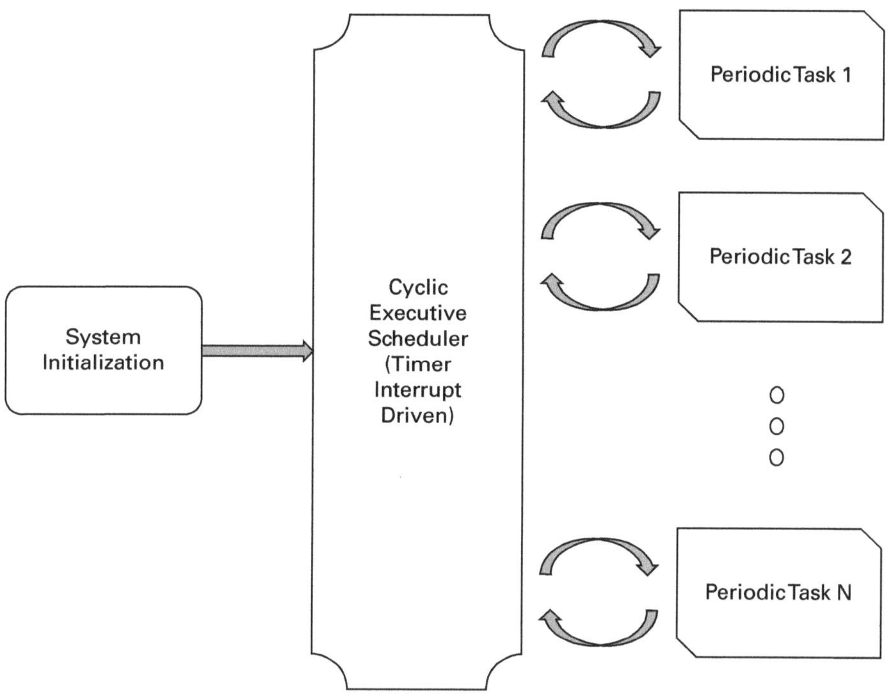

Figure 4.5 Cyclic Executive System.

One of the major differences from ad-hoc scheduling is that task scheduling in a cyclic executive system is periodic rather than repetitive. The tasks are scheduled at a precise periodic rate rather than being based on completion of the prior task in the sequence of tasks as seen in some of the previous *state machine* based approaches. The three major ways in which asynchronous events are handled in cyclic executive systems are: interrupt processing, background processing, or special periodic server tasks. The scheduling of tasks in a cyclic executive system is completely deterministic except for the asynchronous event processing requirements. The two types of timing requirements that have to be considered for cyclic executives are:

- Periodic—tasks that must be performed on a periodic basis
- Deadline—tasks that, when required, must be completed within a certain time interval (the deadline)

A periodic timing requirement will have a deadline that is typically the end of the period or, in some cases, earlier if the task is required to be completed within an interval shorter than the period. The underlying idea is that the cyclic executive design is a loop, or *cycle*, synchronized by timer-driven interrupts. The order in which tasks are scheduled by such a cyclic executive is *predetermined*, with each task having control of the processor (and other resources) for a fixed time period.

Example 4.2.1: A simple cyclic executive schedule.

A simple cyclic executive schedule is depicted in Figure 4.6. The highest rate task, τ_1, executes at 64 Hz. It is scheduled as a result of periodic timer-driven interrupts received by the system. Task τ_2 executes at 32 Hz and is scheduled as a result of every other completion of task τ_1. Task τ_3 is scheduled at 16 Hz, as a result of every other completion of task τ_2. In this simple example, the highest frequency task τ_1 is, therefore, scheduled by the timer-driven interrupt, and the lower rate tasks are based upon the completion of τ_1. It is possible that periodic scheduling of tasks will result in some unused processor time, known as *idle time*. This excess time is typically used for low priority *background processing*. Examples of low priority functions are diagnostics, self checks, and other non-periodic processing.

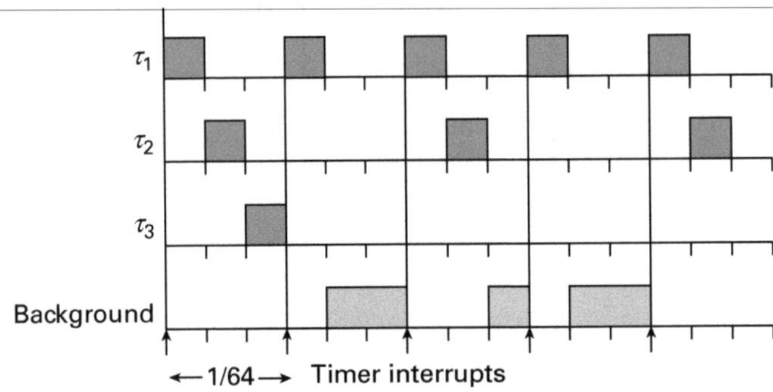

Figure 4.6 A Simple Example of a Cyclic Executive Schedule.

The type of scheduling illustrated in Figure 4.6 is common for cyclic executives, and it produces two kinds of cycles—the *minor cycle* and the *major cycle*. The minor cycle is formed by the repetition of the highest rate task τ_1, and different minor cycles result in the execution of different sets of tasks. The task set executed by a minor cycle is called a *minor frame*. There are four minor frames in the example shown in Figure 4.6. The first frame contains tasks τ_1, τ_2, and τ_3 executed in sequence. The second frame contains task τ_1 followed by background processing. The third frame contains tasks τ_1 and τ_2 followed by background processing, and the fourth frame, again, contains task τ_1 followed by background processing. Following the fourth frame, the entire sequence repeats, producing the periodicity known as the *major cycle*.

The example is a simple, yet effective, illustration of a cyclic executive using timer-driven interrupts to provide the precise periodicity imposed upon the system. It is often the case that asynchronous processing is present in the system, and it can cause severe problems with the synchronous schedule. There are several ways to accommodate asynchronous processing requirements in a cyclic executive design. The most direct way is to have the asynchronous events interrupt the processor using an interrupt service routine (ISR) to perform the required processing.

Figure 4.7 shows the previous cyclic executive schedule with three asynchronous interrupts presenting themselves during the first major cycle. Interrupt processing typically occurs at the highest priority, and each asynchronous interrupt causes the interrupt service routine to immediately preempt the currently executing task. The result is that the tasks are delayed while completing their execution within the minor frame. In the scenario illustrated below, all tasks are delayed but manage to complete by the end of their minor cycle. A serious problem could easily arise with the use of a cyclic executive if the asynchronous interrupts occur too frequently or in particular patterns. The result would be that one or more of the tasks is unable to complete by the end of its minor cycle. Such a condition is known as *frame overrun*.

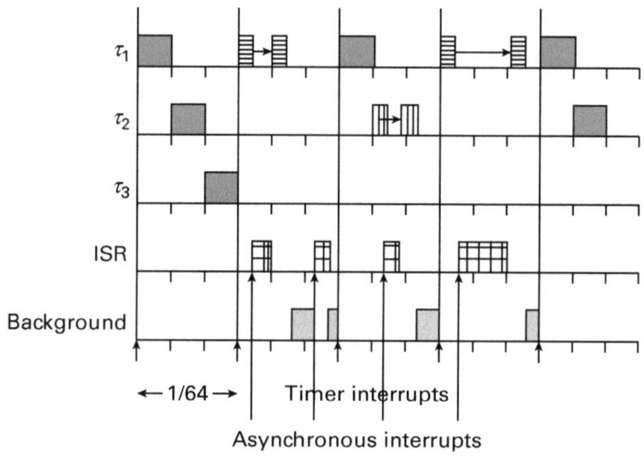

Figure 4.7 The Effect of Asynchronous Processing.

This is an excellent juncture to introduce the concept of *processor utilization* ***U***, especially for the single processor case. The equation below shows how to compute processor utilization for a system with N periodic tasks, and the inequality is a *necessary but not sufficient* condition for a task system to be feasible on a single processor.

$$U = \sum_{n=1}^{N} \frac{e_n}{p_n} \leq 1 \qquad \text{(EQ 4.2.1)}$$

where e_n is the execution time for each task and p_n is the period of each task. This equation does not directly model the effect of asynchronous tasks present in the system. Various approaches may be used to model the effect of aperiodic tasks and are considered later. For the system from Figure 4.6, the processor utilization may be computed as follows (assuming each unit of time on the x-axis is equal to $1/(3*64)$ seconds without loss of generality): $N = number\ of\ tasks = 3$

Period	Deadline	Execution Time
$p_1 = 3$	$d_1 = 3$	$e_1 = 1$
$p_2 = 6$	$d_2 = 6$	$e_2 = 1$
$p_3 = 12$	$d_3 = 12$	$e_3 = 1$

The processor utilization is determined to be $U=\frac{1}{3}+\frac{1}{6}+\frac{1}{12}=0.583<1$. The value can provide an excellent insight into how *stressed* the system is and whether it will meet deadlines consistently. Based on empirical observations, it has been observed that systems at 0.25 or below are *underutilized* and are generally safe below 0.69. Depending on the number of tasks, they are *questionable* till 0.83, *dangerously poised* thereafter until 0.99, *marginally stressed* or *critical* at 1.0, and definitely *overloaded* or *stressed* at any value greater than 1.0.

4.3 The Multitasking Executive and Its Implementation

The more sophisticated approach to the task-scheduling problem is a *multitasking executive* that schedules tasks in real time based on their current priority, which may be chosen based on a scheduling strategy or metric. Such a priority-driven approach does provide a great amount of flexibility since tasks can be added or deleted or the priorities of tasks can be changed without the necessity to change the system or redesign it. The priority-driven multitasking executive is appropriate for systems with

- Synchronous requirements
- Non-deterministic task execution times (times may vary over successive executions)
- Large asynchronous components

A perfect example is a C^3 (command, communication, and control) system in which there are some periodic tasks but which must respond to large numbers of asynchronous events as data are collected and processed by the system.

Figure 4.8 shows a typical multitasking executive (RTOS) based real time system. The basis of priority-driven scheduling is that tasks are assigned priorities and are dynamically scheduled based on those assigned priorities. At any point of time, the task (or tasks if there are multiple processors) with the highest priority will be scheduled for execution unless the execution of the task is blocked. A task is said to be *blocked* if the current state of program execution requires that it not be allowed to execute. For example, the task may be waiting for some lower priority task to provide it with needed data. The core functional component that enables priority-driven scheduling is a software component called a *multitasking executive*.

In addition to task scheduling, the multitasking executive usually provides other services such as task synchronization, intertask communication, and memory management. Processing necessitated by asynchronous events is simply considered to be another task and assigned an appropriate priority. The multitasking executive then schedules asynchronous processing in deference to the assigned priority. All task scheduling is *non-deterministic* since the tasks are dynamically scheduled as the program executes.

Priority-driven scheduling is, thus, accomplished using a *multitasking executive*, which is typically a commercial product due to its complex nature. In contrast, a cyclic executive uses the concept of timer-driven interrupts to produce periodic scheduling of tasks. A multitasking executive schedules tasks *dynamically* (as the program executes) in accordance with the established

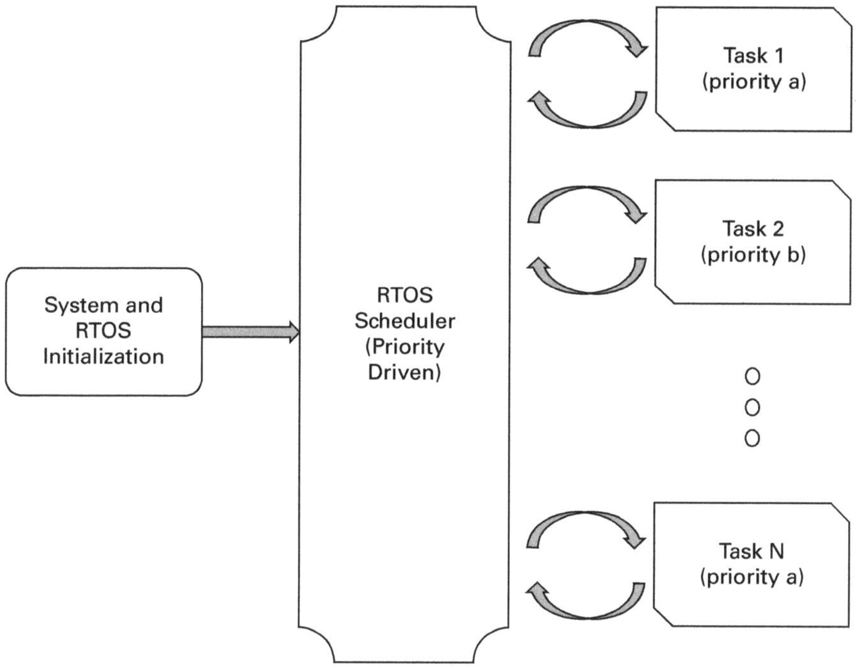

Figure 4.8 Multitasking Executive (RTOS)-Based Systems.

priorities of the various tasks. A cyclic executive accomplishes task synchronization by partitioning each task into subtasks and then deterministically scheduling the subtasks to produce the required synchronization.

In priority-driven systems implemented using a multitasking executive, the scheduling of tasks is non-deterministic and, consequently, so is task synchronization. When tasks are scheduled using a multitasking executive, task synchronization involves suspension and resumption of a task in accordance with the status of other tasks. Various primitive operations, such as semaphores and queues, are provided by the multitasking executive and are used for task synchronization and for communication among tasks.

The suspension and resumption of tasks require certain fundamental operations implemented by the multitasking executive. These operations generally involve the *context* of a task. The context is the minimal information about a task that must be saved before suspension of the task so that it may be resumed at a later instant. At a minimum, the context of a task consists of all or some of the processor register contents at the time of suspension and other information as necessary. The context of a task, together with other information about the task, is maintained by the multitasking executive in a data structure contained in protected memory called the *task control block*, or TCB. When a task is suspended, its context is stored in its TCB. When the task is resumed, the context is read from the TCB, and the processor is restored to the condition existing at the time of suspension.

This process of replacing the context of a suspended task with that of a resuming task is called *context switching*. Context switching takes additional time that is overhead from the viewpoint of the application program. The amount of time required for context switching is one of important benchmarks used to judge the efficiency of a multitasking executive.

It is extremely important to distinguish between *concurrent execution* and *parallel execution,* particularly in relation to the real time systems execution environment. The concurrent execution of multiple tasks is only an impression provided to the user that multiple tasks are running at the same time because a single processor can only implement one task at any instant of time. Multiple tasks can be *active* at any instant of time and execute sequentially using a simple *time-slicing* scheme (also known as *round-robin* scheduling). Alternatively, such context switching can also be based on preemption due to priorities or resources. Many modern real time systems use multicore or multiprocessor configurations and can actually execute tasks in parallel. There will typically be more tasks than processors even on such systems, thereby, resulting in the requirement for concurrent execution of the tasks across multiple processors.

The basic idea behind concurrent execution of tasks is that they do not execute simultaneously in real time but rather time-share common resources needed for execution, with the most critical common resource being the processor itself. The primary function of a multitasking executive is to schedule the tasks for execution in a manner that ensures the meeting of system operational requirements. In this scenario, the tasks are suspended and resumed during execution, but this context switching is totally non-deterministic and, therefore, cannot be predicted by the real time systems developer. This non-determinism is the precise reason that synchronizing mechanisms (e.g., semaphores, signals, etc.) must be provided by the multitasking executive systems software.

The multitasking executive basically performs context switching by storing the necessary registers required to resume the task in a safe location and, then, replacing them with the registers for the task that is taking over. The context of a task, as stated earlier, is the minimum data required to resume a task that has been previously been suspended. At a minimum, the context of a task consists of certain processor registers, such as the program counter and stack pointer. More realistically, many other register contents are involved related to input/output or memory management.

4.4 The Concurrently Executing Task

A multitasking executive or real time operating system (RTOS) must provide for the creation, deletion, preemption, and monitoring of tasks or threads. The term *thread* is increasingly used instead of the term *task* to describe a distinct sequence of operations since the use of multicore processors has become commonplace over the last few years. Throughout this book, the two terms will be used interchangeably. Storage space requirements for the context of a task must be specified, and provision must be made for multiple instances of the same task. Other properties such as the task identifier and the task priority must also be maintained. It is convenient to maintain such information about a task or thread in a data structure or object. Such an object is often known as the *task control block*, or TCB, or Thread Definition Structure (TDS). The specific implementation varies with the operating system but, typically, such a TCB or TDS associated with each task will usually be in a section of memory with restricted access. A possible implementation of the data structure contains a *register save area*, in which the context of the task can be saved with a number of link fields that are used to implement the various functions of the multitasking executive. These link fields may be used to form various linked lists used in task scheduling and in the implementation of operators such as semaphores.

Example 4.4.1: Thread Definition Structure (TDS) and Task Control Block (TCB) examples.

Simple examples of a TDS and a TCB are shown in Figure 4.9. The various fields of a simple TDS are as follows:

- THREAD_ADDRESS—the pointer to the function which implements the thread.
- THREAD_PRIORITY—the initial thread priority which may remain fixed or change during the system operation based on static or dynamic priority assignment.
- INSTANCES—the maximum number of concurrent instances of the thread function allowed.
- STACKSIZE—the stack size requirements in bytes for the thread (0 is default stack size).

The fields of a typical TCB are listed below:

- TASK_NAME—the name of the task that acts as an identifier for starting the task, suspending the task, or performing some other operation affecting the task.
- PRIORITY—the priority of the task which may remain fixed or change during the system operation based on static or dynamic priority assignment.
- STATE—the state of the task, typically—*running* (currently has the necessary resources and is executing); *suspended* (currently blocked from execution awaiting action); or *ready* (not blocked from execution, but waiting for resources necessary for execution). Tasks in the ready state are on the ready list, a linked list of TCBs from which the highest priority task is selected for execution when the currently executing task completes or is suspended.
- ACTIVATION_TIME—the time instance in the future when the task will be activated which consists of placing it on the ready list.
- Register save area—the area in the TCB where the processor register contents are stored upon task suspension. It is from here that the register contents are retrieved when the task is resumed.

TDS – Thread Definition Structure

```
Struct osThreadDef {
Function Pointer *pthread;
Priority level    priority;
Uint32_t          instances;
Uint32_t          stacksize;
};
```

TCB – Task Control Block

```
Struct osTaskControlBlock {
String      TASK_NAME;
Uint32_t    priority;
Uint32_t    STATE;
Time t      ACTIVATION_TIME;
Char *Register Save;
};
```

Figure 4.9 Simple Thread Data Structure and Task Control Block.

For each task that is active, a TDS or TCB is created so that the multitasking executive knows of the existence of the task. Task creation is typically a function of the multitasking executive and usually can be accomplished dynamically as the software executes. Other functions that may be provided by multitasking executives are interprocess or intertask communication using semaphores, mailboxes or messages, interrupt handling, memory allocation, input/output, device drivers, and file support.

4.5 The System Timer

One of the most important tasks for any modern operating system is to maintain accurate time. In this context, time does not necessarily mean the actual world time displayed on the clock but, rather, a measure of elapsed time between events. Typically derived from the system clock frequency, this function is performed by a special timer subsystem that can be set to interrupt the operating system at regular intervals. Typically, this value can range from 1ms to 100ms. In real time systems, the most common value is 10ms. Usually, this value can be changed if more accuracy or granularity is required. The clock tick interrupt can be viewed as the system's heartbeat. The rate is application specific and depends on the desired resolution of this time source. However, the faster the tick rate, the higher the overhead imposed on the system. The timer exception enables the operating system to perform task management and context switching accurately.

The implementation of such a system, also known as the *real time clock*, is supported in hardware by means of a timer peripheral which may be on the chip or off the chip. For example, most of the modern ARM processor cores have an additional component on them known as the *SysTick* timer. Such timers can be implemented as simple increment or decrement counters that produce a hardware interrupt or system exception that must be handled by the CPU. The operating system treats this exception at a high-priority level, and the *scheduler* part of the RTOS is invoked at this time. It makes decisions about system level issues such as which task should run in the next time slice.

Many manufacturers like ARM have introduced the system timer module as a part of their CPU core to allow software portability across their array of product offerings and also for convenience of operation for real time applications. The system timer can also be used for other important tasks like calibration, task time profiling during debugging, and measurement of time between events. Some important issues must be kept in mind while using the system timer. Only the operating system should be allowed to change the system timer. Application programs should be allowed to read the timer value via a protected mode. This functionality for a privileged mode exists on certain processors.

4.6 Preemptive Scheduling

One of the major advantages of using a multitasking executive or RTOS is the ability to stop the execution of a particular task and suspend it until a later time. When such an action is taken intentionally it is known as task *preemption* and is one of the powerful tools that allows the RTOS to meet deadlines imposed by scheduling constraints. Preemption results in the suspension of the currently executing task in order to permit another task to execute. The concept is powerful enough that even though it is applicable in a dynamic run time environment, it can be extended to the design of deterministic schedules by arbitrarily dividing a fixed task into subtasks. The goal, of course, is to facilitate the implementation of an effective task schedule that meets all deadlines.

The simple example shown in Figure 4.10 serves to illustrate the advantage that preemption adds to the scheduling process. Three tasks, τ_1, τ_2, and τ_3, have execution times of 5, 5, and 4 time units and are to be scheduled on two processors. Without preemption, the optimal schedule exhibits an overall execution time of 9, with idle time produced on processor P_2. If preemption is permitted, task τ_2 can be divided into two subtasks, with one subtask executing on each processor, thus, reducing the overall execution time to 7. The advantage is even more apparent with multicore and multithreaded systems.

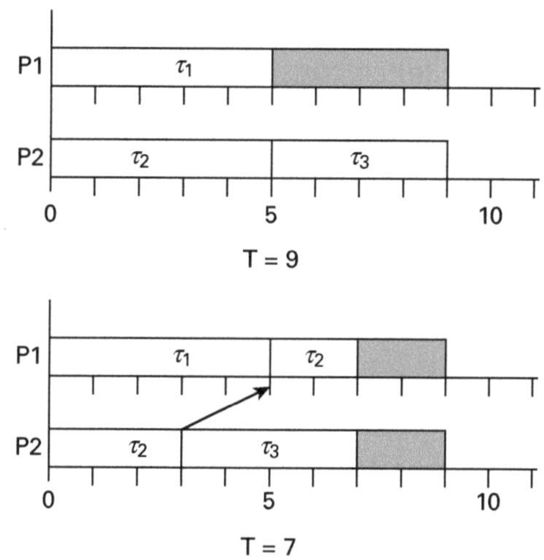

Figure 4.10 Scheduling Advantage Due to Preemption.

Static and dynamic task scheduling algorithms do not have guaranteed closed form solutions under most normal operating conditions and can sometimes lead to unpredictable behavior. This will be clearly observed in the heuristic solutions examples later in the text. Consequently, heuristic algorithms used for scheduling do not always provide optimal solutions. The real time systems developer has to apply additional knowledge from the project to meet system requirements. As systems get larger, with many interdependent parts, it is increasingly difficult to do so, leading to the increasingly prevalent use of dynamic scheduling with techniques based on preemption. It is difficult to apply any dynamic algorithm without allowing preemption and, typically, there is no way idle processor time can be inserted to aid in the scheduling.

The effect of preemption on scheduling algorithms will be examined further in the following chapter since it affects task scheduling at a fundamental level. Inserting processor idle time is relatively straightforward and would probably be the first step in the use of reasoned judgment to adjust a result that seemed unsatisfactory.

The use of preemption has a profound effect on task-scheduling problems for both aperiodic and periodic tasks, treatments for which are considered later in this book.

4.7 Precedence Constraints

A very important factor for any real time system is dependency of one task upon another. Such dependencies are expressed as precedence constraints. An example of a precedence constraint is when one task must complete execution before one or more of the other tasks can begin execution. For practical systems, such precedence constraints occur due to the characteristics of the tasks themselves and may add a layer of complexity to the scheduling process or algorithms. For deterministic schedules, precedence constraints help to satisfy task synchronization requirements. Tasks may be divided into subtasks and with precedence constraints for the subtasks. It is important to characterize such constraints in a formal fashion so as to allow the system designer to develop meaningful algorithms.

The scheduling problem description should now include the following parameters: the number of processors M, the number of tasks N, and the execution times of each of the N tasks e_i and a graphical method to specify the precedence constraints, known as a *precedence graph*. The precedence graph indicates the precedence constraints and the individual task execution times. An example of a precedence graph is shown in Figure 4.11. The symbol τ_i / e_i at a node represents task τ_i with execution time e_i. The precedence relationship, as shown in Figure 4.12, is that task τ_i, with execution time e_i, should be completed before task τ_j, with execution time e_j, can begin executing. The precedence graph of Figure 4.11 specifies that task τ_6 (execution time of 7) cannot begin executing before both τ_1 (execution time of 8) and τ_4 (execution time of 3) have completed execution.

Precedence constraints add to the difficulty of the scheduling process, and the general problem remains intractable in most scenarios. However, for deterministic schedules, certain methods may be used to obtain good results. In multitasking systems, precedence constraints may be expressed using intertask communications schema such as semaphores, mailboxes, signals, and queues.

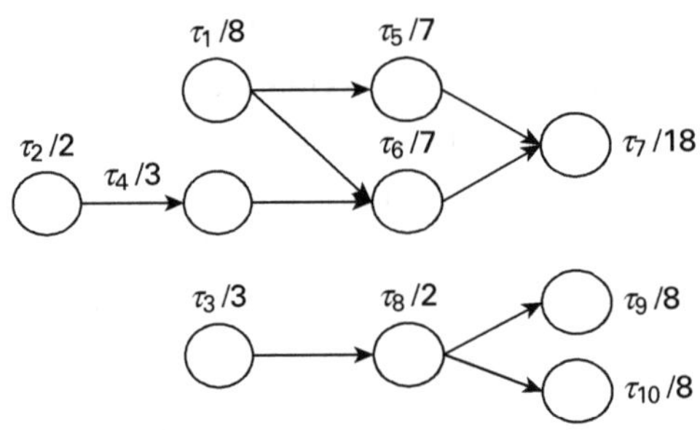

Figure 4.11 An Example of a Precedence Graph Showing Constraints.

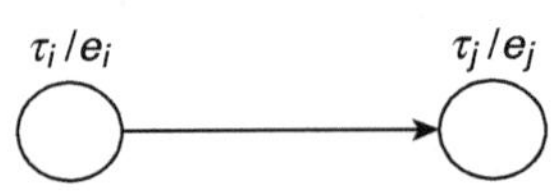

Figure 4.12 Precedence Graph Segment.

4.8 Choosing a RTOS

The multitasking executive described above is the most sophisticated option used to design a real time system and is commercially available as a Real Time Operating System (RTOS). An RTOS has many features including a system timer, supervisor/task modes of operation, task scheduling, task preemption, task precedence settings, intertask communication mechanisms, and other features that make system development very convenient for the designer. The choice of the RTOS is one of the most important decisions made by the system designers. It is very important to review RTOS features before arriving at a choice since it is very difficult to change it once it has been made. This is especially true for developers and companies that must support their software through several generations of the product. Software can be difficult to port and test on a different RTOS in the future.

A number of RTOS alternatives are available in the marketplace and they come in a variety of flavors and pricing options ranging from freeware to royalty-based licenses. Some examples of commercially available multitasking RTOSs are noted in Table 4.1—a list that is, by no means, exhaustive in nature. Among these some notable ones due to their popularity are: ThreadX, Microltron, VxWorks, QNX, RTX, Windows RT, Symbian, RT Linux. Some chip manufacturers also provide free software to entice designers to use their products.

Table 4.1 Commercially Available Real Time Operating Systems—A Variety of Options.

Name	Company/Origin
FreeRTOS.org	Public Domain
POSIX	IEEE Standard
AMX	KADAK
C Executive	JMI Software
RTX	CMX Systems
eCos	Red Hat
INTEGRITY	Green Hills Software
LynxOS	LynuxWorks
µC/OS-II	Micrium
Neutrino	QNX Software Systems
Nucleus	Mentor Graphics
RTOS-32	OnTime Software
OS-9	Microware
OSE	OSE Systems
pSOSystem	Wind River
QNX	QNX Software Systems
Quadros	RTXC
RTEMS	OAR
ThreadX	Express Logic
Linux/RT	TimeSys
VRTX	Mentor Graphics
VxWorks	Wind River
TI-RTOS	Texas Instruments Inc.

It is extremely important to consider the various aspects affecting the choice of an RTOS due to the significant implications it may have on the product meeting specifications. The major items considered while choosing an RTOS are listed below:

- **Cost**—can be an important factor for any commercial system. Some RTOSs are available for no cost, which is quite attractive. For these systems, customer support may be a major issue, and there are no performance guarantees. On the other hand, such distributions provide the source code which may be attractive as the system designers know exactly what is in the code. Modifications of the code may, however, result in a split in the distribution tree, resulting in support issues down the road. The advantage of a commercial RTOS is typically a larger number of features, a guarantee of reliability, and better software and customer support.
- **Speed**—is typically defined in terms of response time or interrupt latency, which are dependent on the internal code and RTOS design. Most manufacturers will guarantee a certain maximum latency for a particular hardware platform. Speed may also be dependent on the run time size—known as the *footprint*—and *scalability* (downward) of the RTOS. Very large run time executables for certain RTOSs may render them useless for applications where the desired footprint is small.
- **Reliability**—has several aspects, including past performance in the field and support mechanisms to prevent deadlock, priority inversion, and other issues related to preemptive tasks. Typically, the longer that a RTOS has been tested in the field in hard real time systems, the better is the expected reliability. Engineers supporting the RTOS product have fixed common issues that may arise in the complex software involved over time.
- **Peripheral Software**—is extremely important in the mobile platforms being developed today that have to support a number of complex and high-speed protocols such as USB, I2C, Ethernet, and other proprietary buses for fast peripherals, such as cameras. If a *software stack* is not available for peripheral hardware, then the system designers may have to spend valuable time and money to develop and test the software, which could distract from actual product development and adversely affect the time-to-market.
- **Portability**—is also important as a RTOS that can be deployed on multiple platforms can make it valuable, to a certain extent. Of more importance is the ability to port the RTOS to future versions of the product being developed that may have to run on more advanced versions of the processor hardware. It is important to review the resources of the company selling the RTOS to ensure that it is capable of adapting to the future in a timely fashion.
- **Development Tools**—can make or break the software team and product as real time debugging in the host-target environment can be challenging even under normal circumstances. Most RTOS manufacturers will have elaborate and sophisticated IDEs with profiling tools to observe task performance and allow developers to ensure that they meet their deadlines. Manufacturers may also provide hardware extensions including JTAG debugging and visibility into the system being debugged. Examples are built-in self test, hardware break points, and watch points.
- **Support**—is extremely important in the fast-paced world of embedded development. In particular, documentation may be sparse with the tight product cycles. It is also extremely valuable to have a knowledgeable and experienced team supporting a product under these circumstances (typically for a higher cost) as opposed to the increasingly common *user group* approach with non-deterministic response times.

The above items should be reviewed by the designers before making any decision. It is quite common for designers to start from system specifications to check if the RTOS will be capable of meeting them and to arrive at a first-cut decision. The next step typically involves a formal rating system of the factors listed above and intensive reference to documentation, customer service, and past experience.

Table 4.2 Rubric to Choose from a Variety of Commercially Available Systems.

FEATURE	CRITICALITY	WEIGHT	RTOS A	RTOS B	RTOS C
Cost—fixed	Low	1.0	0.8	0.7	0.3
Cost—royalty	High	4.0			
Speed—interrupt latency	High	5.0	0.9	0.95	0.7
Speed—context switching	High	4.0			
Footprint size (memory required)	Medium	3.0	xxx	yyy	zzz
Processor support	High	5.0	xxx	yyy	zzz
Compiler	Low	2.0			
Software Libraries	Medium				
Customer Support	Medium	2.0			
Field Testing	High	4.5			
	TOTAL SCORE	**30**	**25.5**	**29.5**	**15.0**

Table 4.2 is an example of a possible rating system that can be used to choose a RTOS if there are a variety of options available. The system engineers can insert values into the table cells based on their research, product specifications, technical documentation, user ratings, and reputation. A systematic study of possible options is critical since real time systems may be in the field for a long period of time and will have to be supported by the manufacturer or customer. It is usually extremely expensive and difficult to justify porting the software for a large system to another platform without extreme market pressure. Such an exercise may result in a loss of perceived system reliability that can only be achieved by operating the real time system in the field under possibly harsh conditions.

CH 5

TASK SCHEDULING ALGORITHMS

Real time systems software is typically designed using distinct deliverables known as tasks or threads. Depending on the type of program, the system will execute the tasks cyclically, periodically, or asynchronously upon demand. As mentioned earlier, real time systems have to meet stringent requirements to ensure successful operation. One of the most important tasks in the design is the fabrication of a schedule in order to determine whether the underlying system can meet the deadline requirements. Due to the very nature of the problem, this estimation can be very difficult, making it important to ensure reserve system capacity and, further, to prevent critical system failure. This chapter covers a variety of real time scheduling algorithms in detail, starting with aperiodic tasks and moving on to periodic tasks on multiple processors.

5.1 Types of Tasks—Aperiodic, Periodic, and Sporadic

The array of task-scheduling problems is broad and varies, depending on the specific application. Some of the characteristics of task systems that affect the process of task scheduling are considered below.

- **Periodicity—aperiodic or periodic:** Most real time systems will have both periodic (synchronous) and aperiodic (asynchronous) system requirements. The ratio of periodic and aperiodic tasks required may vary with the primary requirement being periodic, in some cases, with a small number of asynchronous tasks or vice versa. The proportion may vary, and, in some systems, even if the tasks are repetitive, the scheduling is asynchronous in nature. One example of a system in which the tasks are asynchronous, but not repetitive (typically timer-based), is the personal computer operating system. In this case, user tasks are scheduled only by demand and typically only once. Increasingly, multiprocessor and multicore systems try to schedule tasks evenly across the processors in order to achieve desired throughput such as in a high-speed inspection system. In addition, *sporadic* tasks are defined as aperiodic tasks with a minimum interarrival time between instances and *hard* deadlines, while aperiodic tasks, in many instances, are viewed to be background tasks with *soft* deadlines.

- **Preemption:** Preemptive scheduling refers to the flexibility of suspending one or more tasks to allow another task to execute. Preemptive schedules create more possible scenarios, increasing the gamut of possible solutions but dramatically increase the chances of producing a successful operational schedule. The only drawback is that implementation of preemption needs to be done very carefully as it involves context switching.
- **Precedence constraints or synchronization:** These are typically imposed as a result of resource constraints or sequencing. If one or more tasks must complete execution prior to the execution of other tasks, the complexity of the scheduling problem is increased. If the tasks must synchronize at various points in their execution, scheduling complexity is increased. Intertask communication primitives must be utilized for these purposes.
- **Static or Dynamic:** Schedules for tasks may be predetermined or changed on the fly. The latter is typically only possible using a multitasking executive or RTOS. It the task schedule is predetermined, and the tasks are simply executed according to this fixed schedule, it is known as *static* scheduling. An example of static scheduling is the use of a cyclic executive to schedule periodic tasks. If the tasks are scheduled dynamically at system run time, it is known as *dynamic* scheduling and is heuristic in nature by definition. The tasks are dispatched based on a specific strategy which is usually based on task priorities. An example of dynamic task scheduling is the use of priority-driven scheduling for periodic tasks implemented by a multitasking executive. Static scheduling is deterministic, and dynamic scheduling is inherently non-deterministic.

The next few sections cover techniques for aperiodic and periodic task scheduling. Sporadic task scheduling is covered in detail in Chapter 7.

5.2 Aperiodic Task Scheduling

This section deals with aperiodic task scheduling. Various scenarios applicable to practical systems are considered with examples of techniques used to accomplish task scheduling. While this may appear to be theoretical in nature, the exercises are important to establish and prove analysis techniques and results that may later be extended to the more practical problem of scheduling periodic tasks.

The real time system considered here is assumed to have a fixed number of tasks. The task-scheduling problem is to execute each of the tasks a single time. The cost function is the overall execution time of the entire task set. The goal of the exercises is to schedule the task set on a certain number of processors so as to minimize the cost function. This is the classic problem faced by multiprocessing computer operating systems. In general, the discussion will consider a system of N tasks that are to be executed on a set of M processors.

In order to perform system analysis, it is presumed that the precise execution times of all tasks are known. This is unrealistic in the real world, but such an initial assumption allows problem setup and formalization. The execution of most tasks in the real world is non-deterministic since the precise execution path through the task's code cannot be predicted and may change the execution time in successive cycles. In practice, worst-case execution times or some other stochastic model can be utilized. Additional operational conditions, such as precedence constraints and task preemption, may change the system setup and analysis. The problem statement, therefore, follows:

Minimize the time T required to execute N tasks on M processors

or

Determine if the set of N tasks on M processors can be completed before a deadline D

5.2.1 Non-Preemptive Scheduling with No Precedence Constraints

The scheduling analysis begins by posing the problem in its simplest form with no additional conditions.

- No precedence constraints—Any task may be selected for execution without consequences.
- No preemption—Once a task begins execution, it cannot be interrupted.

Even such a simple formulation has an extremely high complexity, especially if the number of processors M is greater than one, with possible algorithmic solutions to the schedule being computed in non-polynomial time (extremely difficult). This is not practical for the systems designer and, as a result, many heuristic methods to obtain solutions have been developed. Such heuristic methods can produce a solution but not necessarily the best possible solution. In this sense, the solution is known as suboptimal since the schedule created cannot be guaranteed to be optimal even though, in some cases, it may be the best one possible.

The first algorithm to be considered is a heuristic algorithm known as the *Largest Processing Time* algorithm. In this algorithm, the tasks are scheduled on available processors in decreasing order of execution time. This algorithm is also called the *Decreasing-List* algorithm. The tasks must be sorted prior to their scheduling in order to create the order of execution. Typical sorting algorithms have algorithmic computational complexities in the order of $n\log_2 n$, which is considered tractable. The Largest Processing Time algorithm can be stated as follows:

Whenever a processor becomes available, assign to it the remaining task with the largest execution time.

The use of the algorithm is illustrated in the following example shown in Figure 5.1. The task system to which it is applied is specified by:

- M —the number of processors,
- N —the number of tasks, and
- e_i for i ranging from 1 *to* N, the execution times of the N tasks.

Example 5.2.1: Scheduling aperiodic tasks by the Largest Processing Time algorithm producing an optimal result.

The task system is specified as below:

$M = 3 \quad e_1 = 13, \quad e_2 = 8, \quad e_3 = 7, \quad e_4 = 6, \quad e_5 = 4$

$N = 7 \quad e_6 = 2 \quad e_7 = 2$

Applying the algorithm to this task system results in the schedule shown as a task timing diagram in Figure 5.1, in which tasks τ_1 through τ_7 are the seven tasks, and P_1, P_2, and P_3 are the three processors.

The algorithm starts with processor P_1 and assigns the task with the largest execution time, task τ_1. Processors P_2 and P_3 are then assigned the next two tasks in descending order of execution time, τ_2 and τ_3. At this point, task τ_4 is the next task to be assigned a processor, but all three processors are busy and cannot be preempted as per the problem specification. Processor P_3 completes task τ_3 at time $t = 7$ and is the first one to come free. As a result, task τ_4 is assigned to processor P_3 immediately. In a similar manner, task τ_5 is assigned to processor P_2 at time $t = 8$ as is task τ_6 at time $t = 9$. The schedule is completed by the assignment of task τ_7 to processor P_1 at time $t = 13$. The dark shaded areas in the task timing diagram indicate processor idle time. The completion time is the time at which all tasks have completed execution and is recorded as $T = 15$. The order of assignment of tasks τ_6 and τ_7, which have identical execution times of 2, is arbitrary. In this particular case, the Largest Processing Time algorithm has produced an optimal solution since it has a minimum total execution time of 15 which cannot be bettered by any other algorithm under the assumptions made.

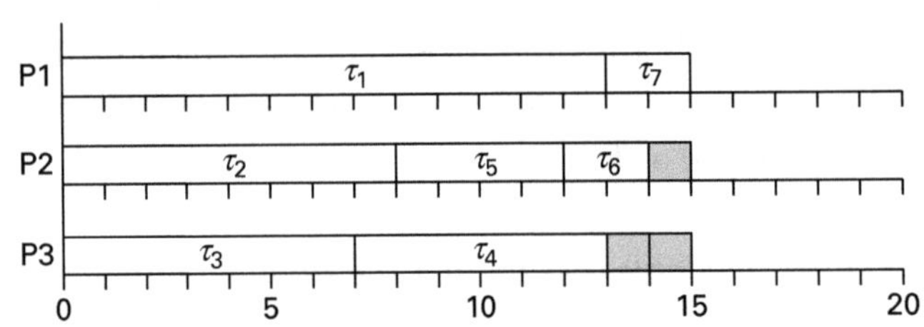

Figure 5.1 Application of the Largest Processing Time Algorithm.

It should be noted that the Largest Processing Time algorithm is not optimal with respect to the scheduling problem being considered here and is a heuristic approach that produced the optimal schedule, largely due to the choice of parameters—number of tasks, processors, and execution times. This is suitably demonstrated by the next example.

Example 5.2.2: Largest Processing Time algorithm can result in a suboptimal result.

The Largest Processing Time algorithm, when applied to non-preemptive aperiodic tasks, can result in a suboptimal solution, as shown in Figure 5.2. The problem is specified as:

$M = 3 \quad e_1 = 16,\ e_2 = 13,\ e_3 = 12,\ e_4 = 8,\ e_5 = 6$

$N = 8 \quad e_6 = 5,\ e_7 = 5,\ e_8 = 2$

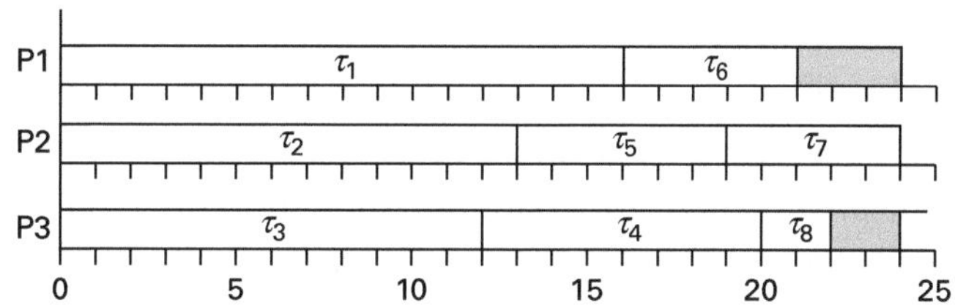

Figure 5.2 Non-Optimal Schedule Using the Largest Processing Time Algorithm.

The total execution time is 24, but, in this case, the result produced by the algorithm is not optimal. A schedule producing the optimal value of 23 is shown in the task timing diagram of Figure 5.3. This example clearly demonstrates that heuristic algorithms can produce good results, and they may even be optimal. However, there is no guarantee that the method will produce the best possible solution and, even more worrisome, is the fact that there is no indication of how good the solution produced by the heuristic algorithm is, i.e., the quality may vary greatly.

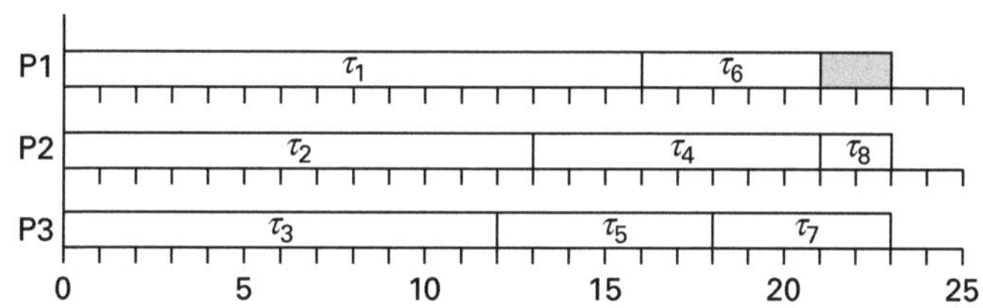

Figure 5.3 An Optimal Schedule Without Using the LPT Algorithm For the System of Figure 5.2.

A performance bound can be determined theoretically for the Largest Processing Time algorithm and is shown in Equation (EQ 5.2.1). For this particular heuristic algorithm, the bound is given as

$$\frac{T}{T_{min}} \leq \frac{4}{3} - \frac{1}{3M} \qquad \text{(EQ 5.2.1)}$$

where M is the number of processors. In the case of the present example, $M = 3$, yields a performance bound of $11/9$. The result shown in Figure 5.2 satisfies this bound. The worst case occurs for large M for which

$$\frac{T}{T_{min}} \leq \frac{4}{3} \qquad \text{(EQ 5.2.2)}$$

This result essentially implies that the worst-case performance of the Largest Processing Time algorithm yields an overall execution time no worse than 33% above the true optimal value. It should be noted that obtaining a truly optimum solution can be very difficult for this particular problem. As a result, the LPT algorithm may be an excellent choice due to the guaranteed performance bound and its simplicity.

5.2.2 Non-Preemptive Scheduling with Precedence Constraints

Precedence constraints arise when one task must complete execution before one or more of the other tasks can begin execution. In practice, precedence may arise naturally from the characteristics of the tasks themselves. They may also result from a requirement for task synchronization, which, in deterministic schedules, is achieved by dividing tasks into subtasks at the synchronization points and establishing equivalent precedence constraints for the subtasks. Thus, the scheduling problem with precedence constraints really involves a number of tasks or subtasks. The further complication of meeting the precedence constraints adds complexity to any heuristic approach.

- A heuristic method known as the *Priority List Scheduling* algorithm is introduced here and described as follows: Establish a list of tasks $\{\tau_1, \tau_2, \cdots, \tau_N\}$ in priority order (specific rules may be based on the application). The list is not required to be consistent with the precedence constraints.
- Assign the highest priority task or subtask for which the precedence constraints have been satisfied to the processor that is currently available or first becomes available.
- If two or more processors compete for a task that is eligible for execution, the tie is broken arbitrarily by assigning the task to the lowest indexed processor.
- If no task is available for a vacant processor, the processor idles until a task becomes available.

A priority list established by decreasing order of execution time, as is used in the Largest Processing Time algorithm for scheduling tasks with *no precedence constraints*, can actually perform quite badly in the presence of precedence constraints.

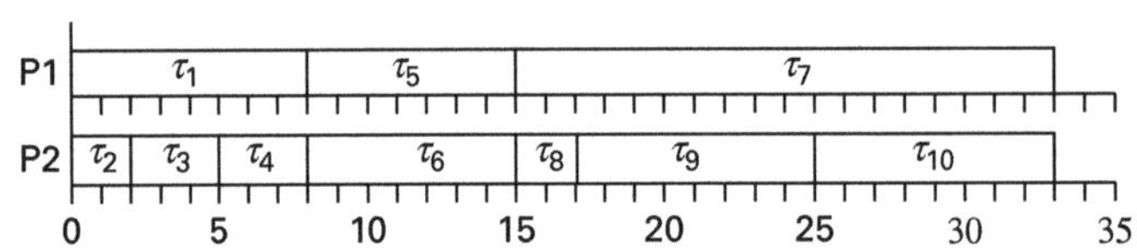

Figure 5.4 An Optimal Solution by Priority List Scheduling.

Example 5.2.3: Priority List Scheduling example produces an optimal result.

The Priority List Scheduling algorithm is illustrated in this example for the set of $M = 10$ tasks on $N = 2$ processors, with the precedence constraint graph specified in Figure 4.11 from the previous chapter. The task system will be scheduled on two processors, using the task priority list

$$\text{Priority list} = \{\tau_1, \tau_2, \tau_3, \tau_4, \tau_5, \tau_6, \tau_7, \tau_8, \tau_9, \tau_{10}\}$$

which is chosen based on the application at hand. Figure 5.4 shows the results of applying the algorithm as described above, with the task timing diagram shown. It can be observed that this is indeed an optimal solution since there is no processor idle time. The minimum overall execution time is 33. In producing this schedule, the algorithm first assigned the highest priority tasks, τ_1 and τ_2, to the processors. Once task τ_2 is completed, processor P_2, is idle, and tasks τ_3 and, then, τ_4 can be scheduled on processor P_2. At time $t = 8$, both processors become idle, and the tie is broken by assigning task τ_5 to processor P_1 and task τ_6 to processor P_2. At time $t = 15$, the process is applied again, and task τ_7 is assigned to processor P_1 and task τ_8 to processor P_2. Tasks τ_9 and τ_{10} are, then, assigned to processor P_2 to complete the schedule.

It is duly noted that for this example, the task list based on the precedence constraints and execution times conveniently permitted each task to be assigned to a processor for execution in strict priority order, allowing a good solution. Such a result is, in fact, quite unusual, and actual results will most definitely be worse in quality and may even be counterintuitive. Decreasing execution times of tasks or increasing the number of processors is not guaranteed to produce a more optimal schedule using the Priority List Scheduling algorithm. It is left as an exercise for the reader to change the number of processors and task execution times, compute new schedules, and compare them with the optimal one.

Two characteristics of the Priority List Scheduling algorithm are responsible for such anomalous behavior: no processor idle time can be inserted to aid in scheduling, and no preemption is allowed. There is a bound on these kinds of anomalous effects when Priority List Scheduling is used:

First, if a schedule is developed for a task system on M processors, with a given set of execution times, precedence constraints, and task priorities, and producing schedule time T; and a second schedule is developed on M' processors with a set of reduced execution times, a weaker set of precedence constraints, and a different priority list, and producing schedule time T', then

$$\frac{T'}{T} \le 1 + \frac{(M-1)}{M'} \qquad \text{(EQ 5.2.3)}$$

This is a strange bound, in essence, limiting the damage that can be done by *lessening* the systems requirements. When $M = M'$, the bound is

$$\frac{T'}{T} \le 2 - \frac{1}{M} \qquad \text{(EQ 5.2.4)}$$

which, of course, approaches 2 as M grows large. There is a corollary to this bound that states: For any set of tasks scheduled on M processors by the Priority List Scheduling algorithm, use of the worst possible priority list can increase the overall execution time by no more than a factor of $\frac{M+1}{M}$ over the optimal value. Here, at least, is a way of evaluating the effects that may occur as a result of different priority lists.

One possible ordering of the priority list is in decreasing order of task execution times (recall the *Largest Processing Time* algorithm). This process is termed *Decreasing Priority List Scheduling*. When precedence constraints are present, this algorithm can create the worst possible schedule bounded by the factor $\left(2-\frac{1}{M}\right)$ (precedence constraints can prohibit tasks from executing in the order specified by the priority list). When there are no precedence constraints, this algorithm exhibits a performance bound of $\left(\frac{4}{3}-\frac{1}{3M}\right)$, which is substantially better than $\left(2-\frac{1}{M}\right)$. This bound implies that the schedule time produced by *Decreasing Priority List Scheduling* will never exceed the optimum value by more than a factor of $1/3$.

5.2.3 Preemptive Scheduling with No Precedence Constraints

Now, consider the case with no precedence constraints but with *preemption* allowed. Under these circumstances, the problem of scheduling N tasks on M processors is tractable, and an algorithm can be designed to provide a solution in polynomial time. An algorithm for producing an optimal schedule with preemption allowed and with no precedence constraints is detailed below:

- Order the tasks in a priority list in the order of decreasing execution times
- Compute the minimum overall execution time T from the expression

$$T = max\left\{max\left\{e_n\right\}, \frac{1}{M}\sum_{n=1}^{N} e_n\right\} \quad \text{(EQ 5.2.5)}$$

  It is important to note that the optimal overall execution time T is determined prior to the development of the schedule itself.
- Starting with the first processor, assign tasks to processors in priority order. If the completion time of the task will exceed T, preempt the task at time T and assign the remaining time to the next processor.
- Continue until all tasks have been assigned. The resulting overall completion time will be T.

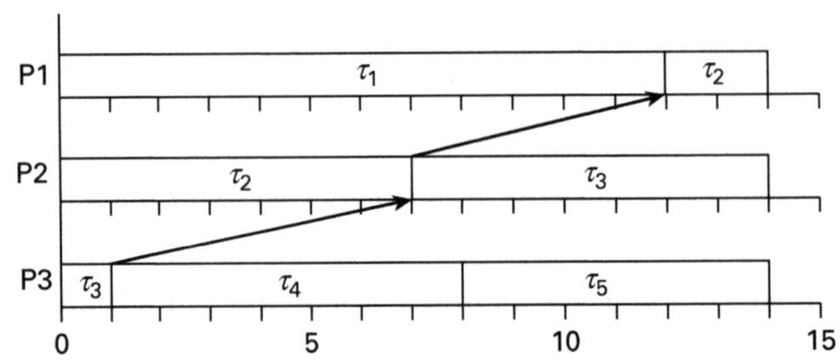

Figure 5.5 Optimum Schedule Using Preemptive Scheduling.

Example 5.2.4: Preemptive scheduling with no precedence constraints.

A simple example serves to illustrate the use of the preemptive scheduling algorithm.

$M = number\ of\ processors = 3$ $N = number\ of\ tasks = 5$

$e_1 = 12, e_2 = 9, e_3 = 8, e_4 = 7, e_5 = 6$

Precedence constraints: None.

The first step is to calculate $T = max\ \{12,14\} = 14$. Then, the tasks are assigned in priority order according to the priority list.

Priority list = $\{\tau_1, \tau_2, \tau_3, \tau_4, \tau_5\}$

The result is shown in Figure 5.5. All tasks have indeed completed by time $T = 14$.

As indicated by the arrows in Figure 5.5, both task τ_2 and task τ_3 were preempted, and the task utilization was 100%. This condition will always be true if the equation below is satisfied:

$$max\{e_n\} < \frac{1}{M}\sum_{n=1}^{N} e_n \qquad \text{(EQ 5.2.6)}$$

The process of creation of the schedule for the task system allocates task τ_2, for example, to be scheduled on processor P_1. The remainder of the task falling outside the total schedule time indicated by the formula is, accordingly, assigned to processor P_2. When the real time system actually executes the tasks, task τ_2 begins execution on processor P_2, is preempted at time $t = 7$, and is resumed on processor P_1 at time $t = 12$.

The use of task preemption is not without its cost. Previous discussion of task preemption has noted that it is accompanied by a requirement for context switching. In the example above, for example, task τ_3's context must be saved at time $t = 1$ and restored on processor P_2 at time $t = 7$. The time required for context switching has not been considered in the discussion above, but it may be a factor in practical application of preemptive task scheduling.

Example 5.2.5: An optimal preemptive schedule with less than 100% processor utilization.

An example of an optimal schedule with processor utilization less than 100% can be produced by increasing the execution time of task τ_1 to 18 in the previous example, which results in a task system that satisfies the inequality of Equation 5.2.4. Now, the overall execution time is $max\{18, 16\} = 18$, and applying the algorithm, then, results in the task schedule shown in Figure. 5.6. It is, of course, assumed that any task cannot be scheduled on two processors concurrently, which is the basis for the optimal time formula in the first place.

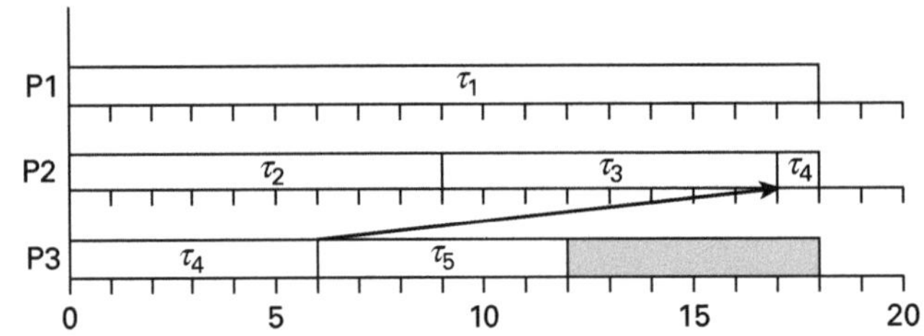

Figure 5.6 Optimum Schedule with Processor Idle Time.

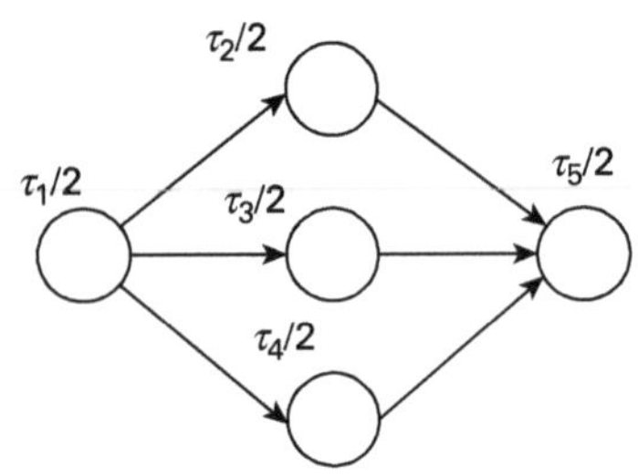

Figure 5.7 Precedence Graph for Processor Sharing Example.

5.2.4 Preemptive Scheduling in the Presence of Precedence Constraints

The concept of *processor sharing* can be extremely helpful in visualizing and creating schedules for tasks executing on multiple processors in the presence of preemption and precedence constraints. Processor sharing considers a processor as a unit of processing capability that can be divided into arbitrary fractions. It can be considered to be an extension of the time-slicing concept on processors in modern operating systems, where each process is assigned a sliver of time on the processor, which is considered a system resource. This approach creates the illusion that the processor is fractionally allocated at any point of time as opposed to the reality that it is a discrete whole. The concept lends itself to be extremely useful when dealing with precedence to create a theoretical processor-sharing schedule, which can be later translated into a realizable preemptive schedule.

Example 5.2.6: The use of processor sharing.

To show the use of processor sharing, a simple problem requiring the scheduling of five tasks with precedence constraints will be considered. The precedence graph is shown in Figure 5.7, in which each task has an identical execution time of 2.

The result of scheduling this task system using the Priority List Scheduling algorithm is depicted in Figure 5.8, with an overall execution time of 8 and processor utilization of 5/8. Now the processor sharing technique will be applied to this example problem.

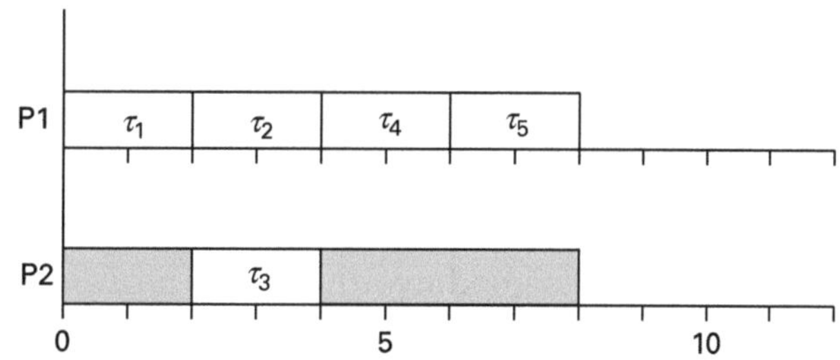

Figure 5.8 Schedule Produced by the Priority List Scheduling Algorithm.

The processor sharing technique consists of assigning any available fraction of a processor's service to any task that is eligible to execute. There is no need for task priorities since all tasks eligible to execute are assigned equal fractions of the total processor capability—the sum of that available on all processors. Tasks execute at a rate proportional to the fraction of a processor assigned to them for execution. For the current example, the processor-sharing schedule is depicted in Figure 5.9. Initially, task τ_1 is the only task eligible to execute and is assigned the entire capability of processor P_1 while processor P_2 remains idle. As task τ_1 completes at time $t = 2$, tasks τ_2, τ_3, and τ_4 all become eligible to execute. In accordance with the concept of the processor sharing technique, equal fractions of the total available processing capability are assigned each of these tasks.

Because the capacity of two processors is available, each task is assigned two-thirds of a processor. Note that task τ_3's share of total processing capability is obtained by assigning one-third of processor P_1 and one-third of processor P_2. At time $t = 4$ these tasks complete, and task τ_5 is assigned to processor P_1, completing at $t = 7$. In the case of task τ_1 at the initial time and task τ_5 at time $t = 5$, a single processor is assigned even though the total capacity of two processors is available. Processor sharing limits the assignment to any one task to the equivalent of one complete processor for reasons that become evident when the conceptual processor-sharing schedule is translated to a realizable preemptive schedule.

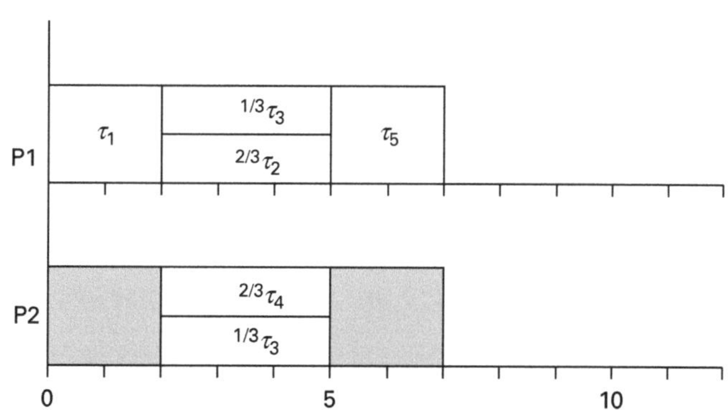

Figure 5.9 Processor-Sharing Schedule.

The translation for this example problem is illustrated in Figure 5.10 and consists of converting time intervals during which processing capacity is shared to equivalent time intervals during which entire processors are time-shared in order to accomplish the required processing. In the current example, there is only a single time interval when processing capacity is shared—the interval from time $t = 2$ to time $t = 5$—during which each task is assigned two-thirds of a processor. The equivalent preemptive schedule assigns tasks to processors in the following manner.

- Task τ_2 is assigned the full capacity of processor P_1 for two time units in place of two-thirds of processor P_1 for three time units.
- Task τ_4 is assigned the full capacity of processor P_2 for two time units in place of two-thirds of processor P_2 for three time units.
- Task τ_3 is assigned the full capacity of processor P_2 for one time unit in place of one-third of processor P_2 for three time units, followed by assignment of the full capacity of processor P_2 for one time unit in place of one-third of processor P_2 for three time units.

The translation is not unique since the assignments to processors can be made in several ways as long as a single task does not execute on two processors simultaneously.

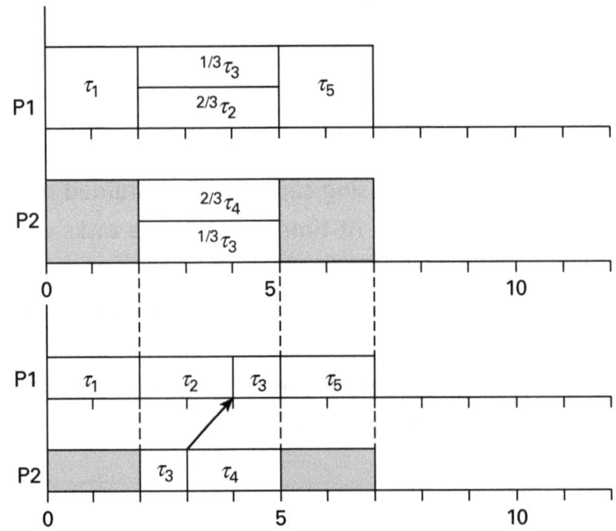

Figure 5.10 Conversion of the Processor-Sharing Schedule to a Realizable Preemptive Schedule.

The general task-scheduling problem in the presence of precedence constraints is generally very difficult. Consequently, solutions are typically obtained by the application of heuristic methods such as processor sharing. It should be noted that there are two specific cases for which polynomial time solutions exist.

1. The number of processors M is one or two.
2. The precedence graph is a tree, and each task has at most one immediate successor (for any number of processors).

For the above situations, a technique called *Critical Path Scheduling* can produce an optimal solution. The Critical Path Scheduling algorithm uses processor sharing and introduces the concept known as *relative urgency*. The relative urgency of a task can change at any given point of time during execution. The various processing chains, headed by the task in question, are identified on the precedence graph. The unexecuted part of the precedence graph is marked along each possible chain, and the sums of the execution times are noted. The relative urgency for the task is then computed as the maximum of these sums of execution times along the possible chains.

This algorithm is designed to ensure that the tasks with the largest values of relative urgency will be executing at any particular time. The name of the algorithm stems from the area of operations research where the largest values of relative urgency are equivalent to a *critical path*. The Critical Path Scheduling algorithm is stated below in a step-wise fashion:

1. Assign tasks to processors according to their relative urgency, starting at the highest level in the precedence graph. If there is a tie among a number of tasks A for the last B processors ($B < A$), then assign the A tasks to the B processors using processor sharing, i.e., assign the B/A of a processor to each task.
2. Reassign the tasks to processors as described above, when either
    - a task is completed, or
    - a time is reached at which one of the tasks is executing at a rate higher than that at which a task of higher relative urgency is executing.

The algorithm ensures that task reassignment occurs continuously in a manner that ensures that the most critical tasks will always be executing.

Example 5.2.7: Use of the Critical Path Scheduling algorithm.

An example of Critical Path Scheduling using two processors is specified by the precedence graph of Figure 5.11. It is not a tree structure. Application of the Critical Path Scheduling algorithm to this problem results in the processor-sharing schedule shown in Figure 5.12. The scheduling process began with assignment of tasks τ_1 and τ_2, both of which have relative urgencies of 8, to processors P_1 and P_2 as indicated below.

- Tasks τ_1 and τ_2 are assigned to processors P_1 and P_2, respectively. As these tasks execute, their relative urgency decreases due to the fact that their execution is shortening the maximum time execution path from their current point of execution to the overall completion time of the task system—their relative urgencies. Eventually, a point is reached at which the relative urgency of these tasks becomes equal to that of one or more of the other tasks in the task system. This point occurs at time $t = 2$.
- At time $t = 2$, tasks τ_1, τ_2, and τ_4 tie with relative urgency of 6. In accordance with the Critical Path Scheduling algorithm, each of these tasks is assigned two-thirds of a processor in the conceptual sense of processor sharing. These tasks now execute at a rate of two-thirds the normal processor rate until time $t = 5$, at which time, each of them accrues two more time units of execution. A task executing for three time units at two-thirds rate will actually execute two time units of code.

- At time $t = 5$, tasks τ_1 and τ_2 complete, making task τ_3 eligible for execution and leaving tasks τ_3 (relative urgency = 4) and τ_4 (whose relative urgency has now decreased to 4 as a result of its execution) tied for the maximum value of relative urgency. Hence, these tasks are scheduled on the two processors.
- At time $t = 7$, tasks τ_3 and τ_4 complete, making the remaining tasks $\tau_5, \tau_6, \tau_7, \tau_8$, and τ_9 eligible for execution, and all five of these tasks tie for relative urgency at the value 2. These tasks are, thus, scheduled—in a processor-sharing sense—on the two processors, executing at a one-fifth rate until completion five time units later at $t = 12$.

The result is the processor-sharing schedule of Figure 5.12. The schedule is obviously optimal since there is no idle time on either processor, a result that is to be expected since the Critical Path Scheduling algorithm is optimal for a two-processor problem. It is sufficient, but not necessary, that an optimal schedule exhibit no processor idle time. Furthermore, the Critical Path Scheduling algorithm is a polynomial time algorithm, making this problem one of the few tractable task-scheduling problems.

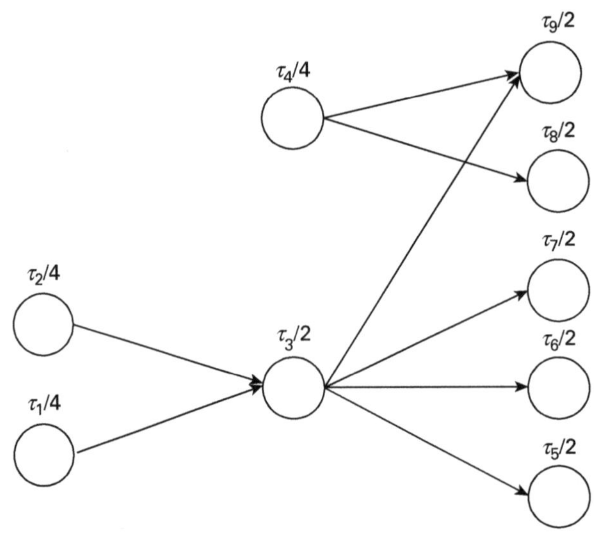

Figure 5.11 Precedence Graph for the Two Processor Example.

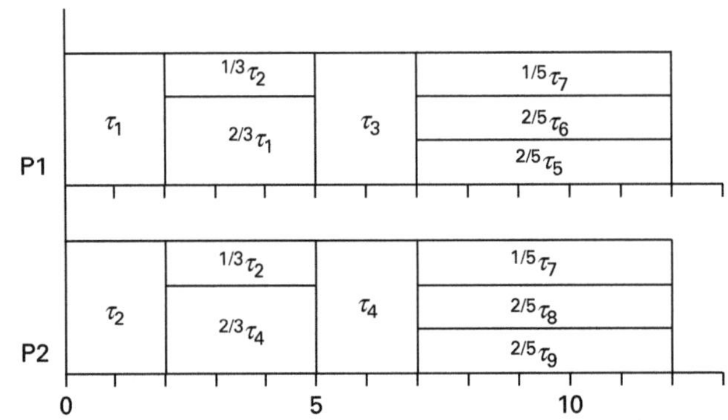

Figure 5.12 Processor-Sharing Schedule for the Two Processor Example.

As discussed earlier, a processor-sharing schedule is not realizable, and the conceptual representation of the schedule must eventually be translated into a realizable preemptive schedule. The process of translating a processor-sharing schedule into a preemptive schedule consists of converting processor sharing over a time interval to the equivalent time-sharing of the processors over the same time interval. The translation of the processor-sharing schedule of Figure 5.12 into a realizable preemptive schedule is shown in Figure 5.13. The process is rather straightforward, requiring only two task preemptions, that of tasks τ_2 and τ_7.

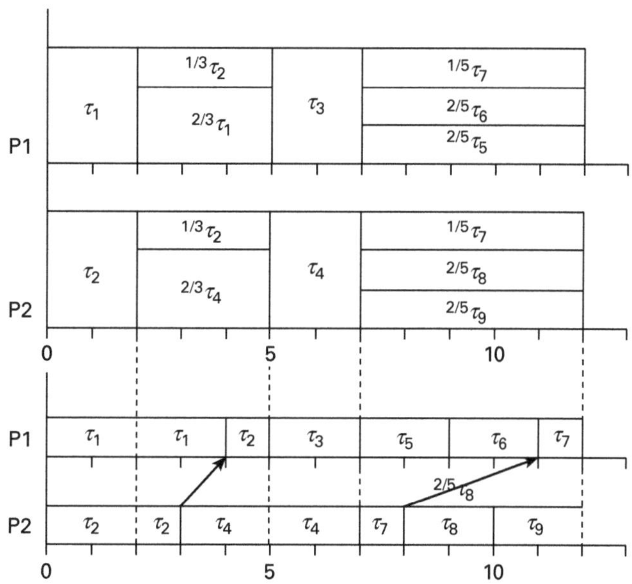

Figure 5.13 Conversion of the Processor-Sharing Schedule to a Realizable Preemptive Schedule.

The preemptive schedule is not unique since, in particular, tasks τ_5, τ_6, τ_8, or τ_9 could replace τ_7 as the preempted task.

As noted previously, the Critical Path Scheduling algorithm is optimal, not only for the special case in which the number of processors is one or two but also when the precedence graph is a tree for any number of processors. The solution of another example problem involving three processors and a tree-like precedence graph is covered next.

Example 5.2.8: Critical Path scheduling with a tree-like precedence graph.

The problem to be considered consists of a task system with nine tasks, with the precedence graph given in Figure 5.14. Task execution times are noted on the precedence graph. The precedence graph is characterized by a tree-like structure, in which each node in the graph has only a single successor node. The number of processors is three. The solution to this problem, produced by application of the Critical Path Scheduling algorithm, is shown in Figure 5.15 in its processor-sharing form. The processor-sharing form of the schedule must then be translated to a realizable preemptive schedule.

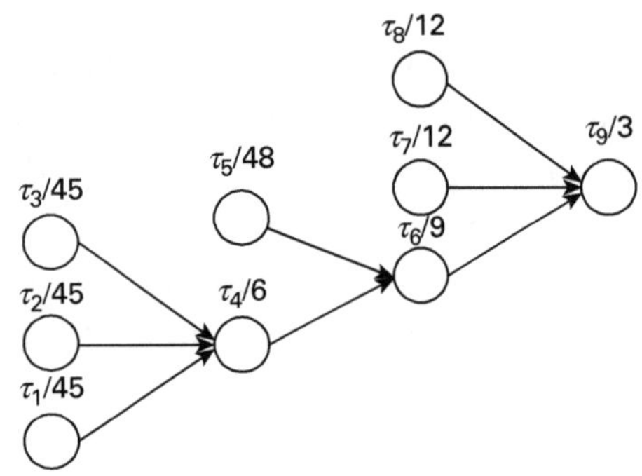

Figure 5.14 Precedence Graph for the Three Processor Example.

Details of the application of the algorithm are given below:

- Initially, tasks τ_1, τ_2, and τ_3 (each with *relative urgency* = 63) are assigned to processors P_1, P_2, and P_3. These tasks will execute until their relative urgencies are reduced to that of one or more of the other tasks.
- At time $t = 3$, task τ_5 ties tasks τ_1, τ_2, and τ_3 with relative urgency of 60. These four tasks then share the three processors at a rate of three-fourths. With a remaining execution time of $t = 42$, the first three tasks execute at the rate of 0.75 per unit time until completion.
- A time $t = 59$, tasks τ_1, τ_2, and τ_3 complete, making task τ_4 eligible for execution, with relative urgency 18. The relative urgency of task τ_5 has been reduced to a value of 18 —as a result of its execution, and it ties with task τ_4. Thus, both tasks τ_4 and τ_5 are assigned full processors— P_1 and P_2. The remaining processor, P_3, is shared by tasks τ_7 and τ_8, which are tied with relative urgencies of 15.
- At time $t = 65$, tasks τ_4 and τ_5 complete, leaving only tasks τ_6, τ_7, and τ_8 eligible for execution. The relative urgency of all three of these tasks is 9, and they are assigned full rate on the three processors.
- At time $t = 74$, tasks τ_6, τ_7, and τ_8 complete, making τ_9 eligible for execution.
- Task τ_9 is scheduled on processor P_1 while the remaining two processors idle.
- At time $t = 77$, task τ_9 completes, and the algorithm terminates, producing the schedule depicted in the upper three lines of Figure 5.15.

The schedule created in Figure 5.15 using the Critical Path Scheduling algorithm is optimal since the precedence graph has a tree-like structure. The processor-sharing schedule not being realizable can be translated to a realizable preemptive schedule. The translation process is shown in the lower section of the figure and is not a unique realization. Note that some tasks have been moved from one processor to another to allow for continuous execution of a task on the same processor.

This section has examined the scheduling of aperiodic tasks on real time systems under different conditions, such as number of processors, preemption, and precedence constraints. The class of tasks termed as *aperiodic* are typically scheduled for execution and are typically a part of a system of similar tasks. The scheduling process results in the development of a task schedule based on a set of tasks with execution times and, possibly, a set of precedence constraints in the form of a precedence graph. The tasks must be implemented on a real time system with multiple processors.

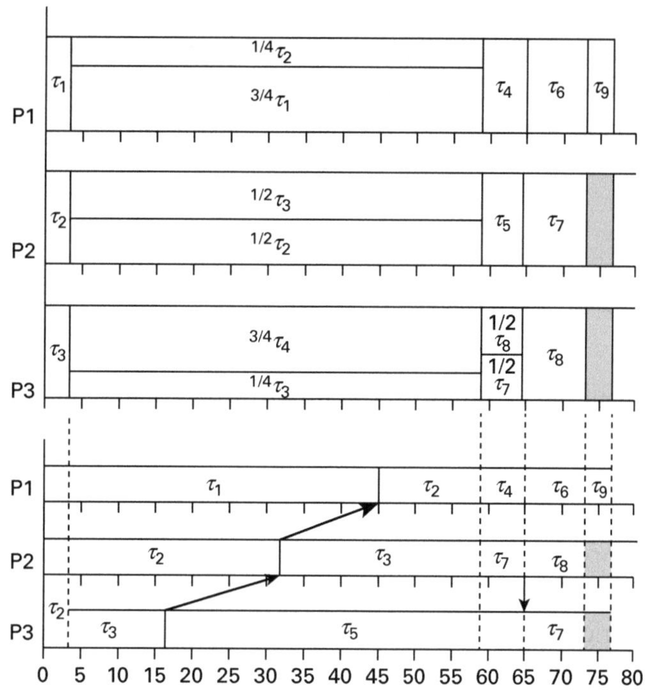

Figure 5.15 Preemptive Schedule for the Three Processor Example.

The goal of the scheduling process is to develop a static schedule that would exhibit minimum completion time so as to achieve maximum efficiency. This poses a problem with a solution that is very difficult to solve, i.e., there is no known polynomial time algorithm that will provide a general solution. It should be noted that the concepts from this section will apply to the single processor case, too, but, typically, the solution will be trivial.

The following basic observations can be summarized based on the analysis presented in this section:

- Task preemption is a critical issue, and it adds additional flexibility to the scheduling process.
- Real time system task-scheduling scenarios involving preemptive tasks may have polynomial time solutions. The Critical Path Scheduling algorithm is an optimal polynomial time algorithm when the number of processors is one or two or when the precedence graph has a tree-like structure with an arbitrary number of processors. Generally speaking, the Critical Path Scheduling algorithm is not optimal, but it may be used as a heuristic algorithm to obtain an answer close to optimal in other situations.
- If the real time task-scheduling problem is very difficult to solve, heuristic algorithms can be and often are used to obtain solutions that may be close to optimal.
- Performance bounds are important tools and can be applied to assess the worst-case performance of heuristic algorithms.
- Heuristic algorithms can and do exhibit paradoxical behavior in the face of certain parameter modifications. Even so, such behavior remains within any established bounds.

The coverage here is not intended to be complete, but, rather, the rudiments of aperiodic task scheduling are presented in order to transition smoothly to the concepts of periodic task scheduling, which will be the topic of the remainder of this chapter. The periodic task-scheduling problem is, in general, more difficult than aperiodic task scheduling. For example, the single processor case in periodic task scheduling is, in general, very difficult, whereas, the aperiodic version of this problem is usually trivial.

5.3 Periodic Task Scheduling

Real time systems typically respond to a number of external or internal stimuli that are frequently periodic. Such periodic requirements may be required for a number of reasons, such as input or output data rates necessary to meet system requirements. As a result, the development of real time systems software is often centered around sets of periodic tasks, and, hence, the periodic task-scheduling model becomes an important issue. Real time systems must also concurrently deal with asynchronous requirements which are defined as demands or tasks that may occur repetitively but not with any specific period or even prior warning. By definition, such processing is not synchronized with the internal system clock or timers, and it is impossible to predict when the demand may arise. The scheduling of periodic tasks in the presence of asynchronous processing requirements is a key issue.

There are several methods used to model periodic task sets that have been established, depending upon real time system requirements placed on them by the designers. These considerations must be recognized and formalized as system parameters:

- **Task Periods:** Periodic tasks have *periods* which are defined as the time interval after which it must be repeated. Task periods are determined by the system requirements and may vary based upon the task specifics.
- **Precedence and/or synchronization constraints:** Tasks do not typically execute independently of one another, and one task may generate data used by another task or wait for it to complete.
- **Task Deadlines:** The time by which task execution must be completed is termed a *deadline*. Usually, a periodic task must be completed by the time it is, again, scheduled for execution (deadline = period), but a task may sometimes have a deadline shorter than the period.
- **Task execution times:** The amount of time a task requires to complete is the *execution time,* and knowledge of these is required in order to develop a schedule. The execution time of a task may vary from one execution to the next. This is known as *jitter* and is due to differing execution paths taken through the code. Jitter can constitute a major problem in periodic task schedules, notably, those implemented by deterministic, timer-driven, cyclic executive software.

It is essential to have an explicit model of just what constitutes a periodic task and to define the properties of that task. The model to be used here is depicted in Figure 5.16. The terminology to be used and its precise meaning are given in the discussion below.

- Task Number: The *n*-th task is referenced as τ_n.
- Release Time: The *release time* of the *n*-th task is labeled r_n. The release time is the time at which a task is scheduled to execute. It may begin execution time any time after its release time, but

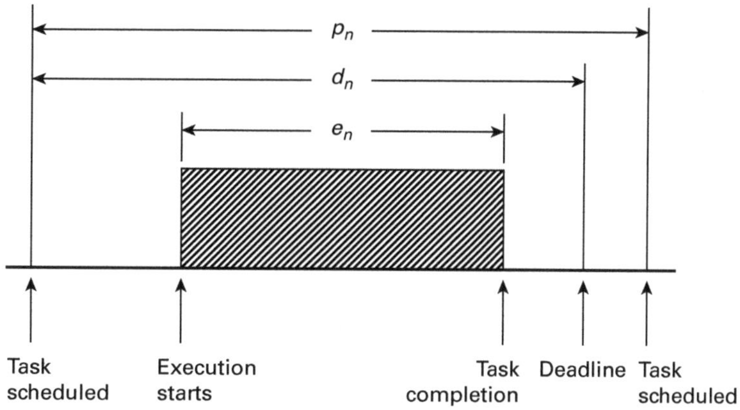

Figure 5.16 Model of a Periodic Task.

it may not begin execution *before* its release time. Release times apply to both periodic and aperiodic tasks, with the release times of periodic tasks occurring periodically.

- Period: The *period* of the *n*-th task is $\boldsymbol{p_n}$, and it is the time interval between successive release times of the task.
- Deadline: The *deadline* of the *n*-th task is designated as $\boldsymbol{d_n}$. The deadline is a time period following the release time of a task within which the task must execute to completion. It is important to note that the deadline references the *release time* of the task, not the time the task actually begins execution.
- Execution Time: The execution time of the *n*-th task is $\boldsymbol{e_n}$. This is the time during which the task is actually executing. The execution time may not begin prior to the release time, nor may it extend past the deadline without serious consequences (missing the deadline).

The above task model does not require the actual execution of the task to be periodic but, rather, that it is the release times that occur in a periodic manner. The periodic task-scheduling problem is to fit the actual task executions into the time periods between periodically occurring release times and their associated deadlines.

In order to meet the real time system-timing requirements, several choices about task scheduling must be made, and these choices affect the scheduling process.

- **Static or dynamic scheduling:** Static schedules are ones that that do not change during real time operation. They are generally implemented using a cyclic executive design driven by an interrupt timer. On the other hand, dynamic schedules may change during system operation and are implemented using a multitasking executive (RTOS) and priority-driven scheduling.
- **Preemption:** Tasks may or may not be allowed to be preempted. For designs based on cyclic executives, preemption means that the tasks can be arbitrarily divided into subtasks to achieve synchronization or in order to *fit* them into the schedule. For priority-driven systems, preemption implies that the real time operating system is free to arbitrarily suspend an executing subtask in order for a higher priority subtask to execute.
- **Asynchronous processing:** Most systems will have requirements represented by aperiodic tasks. The arrival of aperiodic tasks is usually signaled and handled by an interrupt, but normal interrupt processing may not always be sufficient and special server techniques must be applied.

If the task scheduling is preemptive, then the task execution may occur in the form of disjoint units of execution, as depicted in Figure 5.17. The task's execution is composed of k individual periods of execution such that

$$e_n = \sum_{i}^{k} e_{nk} \quad \text{(EQ 5.3.1)}$$

This kind of task execution may occur as a result of the development of a cyclic executive schedule in which the task is broken into subtasks in order to fit it into the schedule. Alternatively, it may occur dynamically as a RTOS preempts and resumes the task during execution. In either case, the task cannot begin execution prior to its release time, and it must complete prior to its deadline.

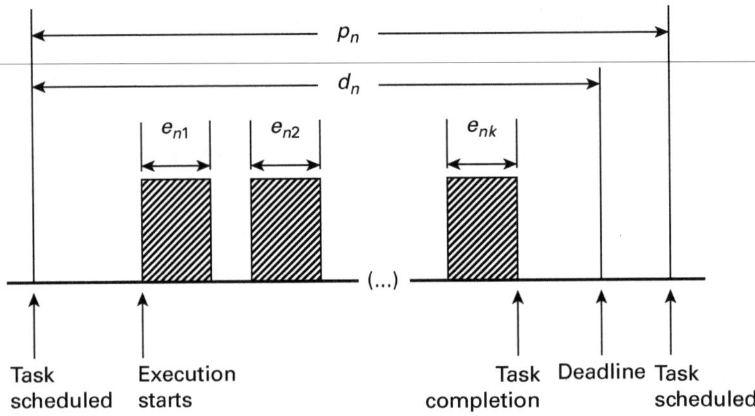

Figure 5.17 Model of a Preemptive Periodic Task.

Task-scheduling problems on real world systems typically deal with the kind of situation depicted in Figure 5.17. On the one hand, the developer of cyclic executive based software must be concerned with dividing tasks into subtasks in order to satisfy synchronization requirements and, then, may well have to further divide those subtasks in order to fit them into the schedule. On the other hand, the developer of software running under a real time operating system must ensure that the task system *is capable of* being dynamically scheduled in a manner that will ensure that all tasks meet their deadlines. Preemptive dynamic scheduling results in just the kind of task execution illustrated in Figure 5.17. Both static (cyclic executive) and dynamic (RTOS) task scheduling will be covered in the text. One of the more common approaches not based on a commercial RTOS is the *cyclic executive*—the non-preemptive static scheduling of periodic tasks on one processor. Such an approach could be addressed here but is deferred to the last chapter on additional topics since it also fits in with coverage on aperiodic and sporadic task handling in the same context.

5.3.1 Preemptive Scheduling of Periodic Tasks

In the previous coverage of aperiodic task scheduling, it was duly noted that arbitrary preemption greatly increases the flexibility available for schedule design. The same effect is noted in the scheduling of periodic tasks. When arbitrary task preemption is allowed, the distinction between the design of cyclic schedules and the schedulability analysis of priority-driven, dynamically scheduled systems tends to blur for reasons outlined below.

1. Synchronization requirements in cyclic scheduling are met by dividing the tasks into subtasks at the synchronization points. Without arbitrary preemption, these subtasks must run to completion once they are scheduled. Preemption adds the flexibility of dividing the tasks into subtasks for the sole purpose of fitting them into the schedule. Even with this added flexibility, however, the general scheduling problem remains very difficult. Consequently, heuristic methods are often used.
2. An effective heuristic design method for generating cyclic schedules is to assign task priorities according to some algorithm and, then, use these priorities to generate the schedule.
3. Dynamic, priority-driven scheduling requires that tasks be arbitrarily preemptable. The primary design question, in this case, is whether or not a given task system with a specified set of task priorities can be scheduled by a multitasking executive in a manner that ensures that all tasks meet their deadlines. Synchronization requirements in dynamically scheduled systems are implemented through the use of semaphores or other primitive operations provided by the multitasking executive.
4. One method of performing the schedulability analysis of a dynamic, priority-driven, task-scheduling process is to build the schedule that would result if all tasks executed for precisely the execution time specified for the task.
5. The result produced in (2) above is precisely the result produced in (4) if the algorithm for assigning the task priorities is the same.

The conclusion stated in (5) above implies that there is no difference in the production of a cyclic schedule using a given set of task priorities, and the verification, through schedule construction, that a dynamic, priority-driven system using a given set of task priorities can meet all task deadlines. For this reason, much of the discussion applies both to cyclic executive schedule design and to the schedulability analysis of dynamic, priority-driven systems.

To summarize the issues discussed above, and to define some basic characteristics of scheduling problems to be considered, the following distinctions are made.

- Dynamic task scheduling refers to the situation in which tasks are scheduled dynamically by a multitasking executive as the program executes.

  Dynamic task scheduling with *static task priorities* refers to the situation in which the scheduler of a multitasking executive dynamically schedules tasks in accordance with a set of *static* task priorities. Static priorities, by definition, are preset and remain fixed throughout the execution of the program.

  Dynamic task scheduling with *dynamic task priorities* refers to the situation in which the scheduler of a multitasking executive dynamically schedules tasks in accordance with a set of *dynamic* task priorities. Dynamic priorities, by definition, can be modified as the program executes. The manner in which these modifications takes place is defined by the particular priority assignment algorithm.
- Static task scheduling refers to the situation in which a cyclic schedule is predetermined, and the scheduler, in this case, a cyclic executive, executes the tasks according to this schedule.

Static task scheduling with *static task priorities* refers to the situation in which a predetermined schedule is developed using an algorithm that assigns priorities to the tasks and then assigns tasks according to these priorities. The priorities are static in that their values do not vary during the development of the schedule.

Static task scheduling with *dynamic task priorities* refers to the situation in which a predetermined schedule is developed using an algorithm that assigns priorities to the tasks and then schedules tasks according to these priorities. The priorities are *dynamic* in that their values vary during the development of the schedule in accordance with the algorithm used for their determination.

The arguments presented above imply that it does not matter whether the goal is to analyze the schedulability of dynamically scheduled, priority-driven systems using static task priorities or to design a static cyclic schedule using an algorithm that assigns tasks according to a set of static priorities. The process is the same. A similar statement holds for the case of dynamic task priorities.

In the discussion of schedulability analysis in the following section, and in the discussion of various priority assignments: deadline-monotonic, rate monotonic, and earliest deadline, which are to follow, no distinction needs be made between the static and dynamic scheduling problems.

5.3.2 Schedulability Analysis of Synchronous Task Systems with Static Priorities

The term *schedulability analysis* refers to those techniques used to determine if a given task system can be scheduled on one or more processors. The discussion of schedulability analysis to follow requires an understanding of certain basic terms and concepts as outlined below.

- A task system is termed ***synchronous*** if the initial release times of all tasks are identical. Without loss of generality, this common release time can be taken to be zero. A task system that is not synchronous is said to be ***asynchronous***. The complexity of a scheduling problem varies considerably depending on whether the task system is synchronous or asynchronous.
- A schedule is said to be ***valid*** for a specific task system if the schedule provides for meeting the deadlines for all task requests. The term ***all task requests*** includes the infinity of requests that exist for a periodic task, and the question that is immediately raised concerns the problem of guaranteeing deadlines for an infinite time.
- A task system is said to be ***feasible*** on M identical processors if there is a *valid* schedule for the task system on M identical processors.
- A task system is said to be ***schedulable*** on M identical processors if there is a *valid* schedule on M identical processors produced by a *static* task priority assignment. The priority assignment may be used by a multitasking executive to dynamically schedule the tasks, or it may be used as a scheduling algorithm in developing a static task schedule.
- A scheduling algorithm is said to be ***optimal*** if it always produces a *valid* schedule for every task system that is *feasible*. This use of the term *optimal* deserves particular attention. It implies that if a task system is feasible, an optimal algorithm will produce a valid—but not necessarily uniquely so—task schedule.

Before proceeding with the development of schedulability tests, two task system properties and their significance should be noted. The concept of *processor utilization* U, generalized to the M processor case, produces the following inequality as a *necessary but not sufficient* condition for a task system to be feasible on M identical processors, and it follows directly from the single processor case.

$$U = \sum_{n=1}^{N} \frac{e_n}{p_n} \leq M \qquad \text{(EQ 5.3.2)}$$

The concept of *deadline utilization* D leads to the following inequality as a *sufficient but not necessary* condition for a task system to be feasible on M identical processors.

$$D = \sum_{n=1}^{N} \frac{e_n}{d_n} \leq M \qquad \text{(EQ 5.3.3)}$$

For synchronous task systems in which all task deadlines are equal to the corresponding periods, i.e., $d_n = p_n$ for all τ_n, the inequality of the first equation above is both necessary and sufficient. In the general case, however, the problem of deciding whether an arbitrary task system is feasible on M processors remains very difficult for all M greater than one. Even for the situation in which the inequality of processor utilization equation provides an easy answer to the YES/NO question, is the task system feasible?, the problem of actually developing the schedule may remain difficult.

5.3.3 Scheduling Synchronous Task Systems with Static Priorities on One Processor

The critically important objective for a systems designer is to confirm that no real time constraints are violated. This can only be achieved if there are some means to confirm that all tasks on the system will meet deadlines for all release times. In technical terms, one needs to determine whether a schedule produced by a particular algorithm for *static* priority assignment is *valid*. Generally speaking, this is a hard problem unless additional constraints are available. It so happens that for task scheduling on a *single processor* there is a pseudo-polynomial time algorithm for determining if the task schedule produced by a given set of static task priorities is valid. At this point, to facilitate further technical discussion, the concepts of task response time and a critical instant are defined.

- The *response time* of a task is the time interval between the release time of the task and the completion time of the task. The response time is different from the execution time of the task. While the execution time is the total of the actual time the task spends executing, the response time is comprised of the actual execution time plus any suspension time.
- A *critical instant* of a task is a time at which the task is released and has the largest response time of all task releases.

The *critical instant* concept necessarily defines the extreme condition under which a task will meet its deadline under the worst-case scenario. It is easy to visualize that the task deadline will be met *for all release times* of the task *if and only if* the deadline is met when the release time occurs at a critical instant. Additionally, this allows the designer to pose situations and determine possible critical instants. Under *static* task priorities, a critical instant will occur for a particular task if it is released simultaneously with all higher priority tasks in the system. Furthermore, if the task system is *synchronous,* the initial releases of all the tasks occur at $t = 0$. This simplifies analysis since a critical instant for all tasks occurs at time $t = 0$.

One must ultimately find an algorithm or method to answer the YES/NO question, is the schedule produced by the given priority assignment valid? For the case developed so far, this algorithm is simply to construct the schedule through the end of the *longest task period*. Being the worst case, if all tasks meet their first deadline, they will meet their deadlines for all future release times. The schedule construction is required only through the longest task period rather than through the entire major cycle. This is the reason for the pseudo-polynomial time characteristic of the algorithm. The following important result can be stated:

For scheduling a synchronous task system on a single processor, there is a pseudo-polynomial time algorithm for deciding if the schedule produced by a given static priority assignment is valid. The algorithm is the construction of the schedule from a critical instant through the end of the longest period.

Instead of the graphical construction of the schedule, it is sometimes more convenient to have analytical techniques to investigate task system schedulability on a single processor. This is particularly true for larger complex systems, and, fortunately, such algorithms do exist. There is one major caveat about the task execution times used in schedulability analysis algorithms. In real world systems, the actual execution times may change from one release to the next. Such statistical variation is caused by jitter and other random phenomena. The two general analysis approaches to allow distinction between the ideal execution time and the actual execution time associated with a particular release time are:

- The calculation of a utilization bound using worst-case analysis. These bounds represent sufficient but not necessary conditions for a task to meet its first deadline.
- Analytical equivalent of schedule construction involving the solution of an iterative equation for the first completion time of the task. If the first completion time does not exceed the associated deadline, then the task will meet its first deadline.

To summarize, in the design of a cyclic executive schedule, it is the construction of the schedule that is of interest. In the analysis of a dynamically scheduled, priority-driven system, it is the answer to the YES/NO question that is of interest. For synchronous task systems to be scheduled on one processor using static priorities, if a task meets its first deadline, it will meet its deadlines for all releases.

For synchronous task systems to be scheduled on one processor using static priorities, if a task meets its first deadline, it will meet its deadlines for all releases. For reasons that will become apparent, the utilization bound method will be addressed first.

5.3.4 Utilization Bound for Scheduling Synchronous Task Systems with Static Priorities

Analytical methods such as utilization bounds are sometimes more convenient to determine schedulability of a task system than creating a graphic schedule. The response times of the tasks in the system must first be determined under any specific set of task priorities. Further, a method to determine the worst-case task response time must be formulated and is typically composed of the three components below:

- The *execution time* of the task itself is the actual time the task spends executing its code.
- The *preemption time* is the time the task is suspended because of preemption by *higher* priority tasks.
- The *blocking time* is the time the task is blocked by *lower* priority tasks.

Blocking can be very important for schedulability analysis and occurs when a low priority task prevents a high priority task from executing because it is blocking a system resource required by both tasks. Other forms of blocking involve a phenomenon known as *transitive blocking,* wherein a high priority task is indirectly blocked by a medium priority task that has preempted the guilty lower priority task holding the critical resource. Blocking times are typically assessed as worst-case values because they are non-deterministic.

The concept of processor utilization that, hitherto, was based solely on task execution time has to be extended to include some additional effects:

1. Preemption by tasks with priorities *greater than or equal to* that of task τ_n but with periods *greater than or equal to* the deadline of task τ_n. The set of tasks with priorities higher than that of task τ_n is designated as S_n. The subset of S_n containing those tasks with periods greater than or equal to the deadline of task τ_n is designated as S_{ng} with the number of tasks in the subset being N_{ng}. Tasks in subset S_{ng} can preempt task τ_n *only once* before its deadline.
2. Preemption by tasks with priorities *greater than or equal to* that of task τ_n but with periods *less than* the deadline of task τ_n. The subset of S_n containing those tasks with periods less than the deadline of task τ_n is designated as S_{nl} with the number of tasks in the subset being N_{nl}. Tasks in subset S_{nl} can preempt task τ_n *multiple times* before its deadline.
3. Blocking delays produced by *lower* priority tasks. Blocking due to any cause is included in the value of worst-case blocking time b_n.

The effective task utilization based on all these possible delays can now be defined as:

$$E_n = \sum_{S_{nl}} \frac{e_i}{p_i} + \frac{1}{p_n}\left[e_n + b_n + \sum_{S_{ng}} e_i \right] \qquad \text{(EQ 5.3.4)}$$

Using this equation, the effective utilization of each task τ_n is computed and compared to a worst-case utilization bound. If E_n is less than or equal to the utilization bound value, the task will meet its first deadline and, therefore, all future deadlines. Each task in the system must satisfy this test for a comprehensive conclusion. On the other hand, if the effective utilization of a task exceeds the bound, no information is conveyed. That is, a worst-case bound, by definition, is a *sufficient but not necessary* condition. In the event that the utilization bound test is inconclusive, the analysis must resort to schedule building or alternative equivalent technique.

The worst-case utilization bound for task τ_n is determined as follows:

$$UB_n = \begin{cases} \dfrac{d_n}{p_n}, & \dfrac{d_n}{p_n} \le \dfrac{1}{2} \\ \left[N_{nl}+1\right]\left[\left\{\left(\dfrac{2d_n}{p_n}\right)^{\frac{1}{N_{nl}+1}} - 1\right\} + 1 - \dfrac{d_n}{p_n}\right], & \dfrac{d_n}{p_n} > \dfrac{1}{2} \end{cases} \qquad \text{(EQ 5.3.5)}$$

If $d_n = p_n$, the equation reduces to

$$UB_n = \left(N_{nl}+1\right)\left(2^{\frac{1}{N_{nl}+1}} - 1\right) \qquad \text{(EQ 5.3.6)}$$

Example 5.3.1: Using the utilization bound test.

The utilization bound test can be applied to the following task system.

N=number of tasks=5

M = number of processors = 1

Period	Deadline	Execution Time	Blocking Time
$p_1 = 8$	$d_1 = 2$	$e_1 = 1$	$b_1 = 0$
$p_2 = 60$	$d_2 = 60$	$e_2 = 16$	$b_2 = 0$
$p_3 = 36$	$d_3 = 28$	$e_3 = 4$	$b_3 = 0$
$p_4 = 50$	$d_4 = 30$	$e_4 = 2$	$b_4 = 1$
$p_5 = 30$	$d_5 = 30$	$e_5 = 2$	$b_5 = 0$

Priority list = $\{\tau_1, \tau_2, \tau_3, \tau_4, \tau_5\}$

The utilization bound test is required to be applied to each task separately.

Test task τ_5:

Determine $S_{5g} = \{\tau_2, \tau_3, \tau_4\}$. This is the set of tasks with priorities greater than or equal to that of task τ_5 and with periods greater than or equal to the deadline of task τ_5. Note that $p_1 = 8$ is less than $d_5 = 30$, thereby, excluding task τ_1 from this set. The number of tasks in this set is $N_{5g} = 3$.

Determine $S_{5l} = \{\tau_1\}$. This is the set of tasks with priorities greater than or equal to that of task τ_5 and with periods less than the deadline of task τ_5. The number of tasks in this set is $N_{5l} = 1$.

Calculate $\frac{d_5}{p_5} = 1$ and use EQ 5.3.5 to calculate the utilization bound. Since $d_5 = p_5$, we can use the simplified equation

$$UB_5 = (1+1)\left(2^{\frac{1}{1+1}} - 1\right) = 0.828$$

This value will be compared to the effective utilization of task τ_5 calculated as follows:

$$E_5 = \sum_{S_{5l}} \frac{e_i}{p_i} + \frac{1}{p_5}\left[e_5 + b_5 + \sum_{S_{5g}} e_i\right]$$

$$E_5 = \frac{1}{8} + \frac{1}{30}[2+16+4+2] = 0.925$$

The effective utilization of task τ_5 is larger than the utilization bound. This result implies only that the utilization bound test cannot conclusively determine if task τ_5 will meet its first deadline. It does not imply that task τ_5 will miss its first deadline.

Obviously, some other test must be applied if a determination of the schedulability of task τ_5 is to be made. Such a test will be described in the following example, but, first, the utilization bound test will be applied to the remaining tasks.

Test task τ_4:

$S_{4g} = \{\tau_2, \tau_3\}$ $\quad N_{4g} = 2$ $\qquad$ (Note: Tasks with periods greater than $d_4 = 30$)

$S_{4l} = \{\tau_1\}$ $\quad N_{4l} = 1$

$$UB_4 = 2[2(0.6^{0.5} - 1) + 1 - 0.6] = 0.591$$

$$E_4 = \frac{1}{8} + \frac{1}{50}[2+1+16+4] = 0.585$$

Since the effective utilization is less that the utilization bound, task τ_4 will meet its first deadline.

Test task τ_3:

$S_{3g} = \{\tau_2\}$ $N_{3g} = 1$

$S_{3l} = \{\tau_1\}$ $N_{3l} = 1$

$$UB_3 = 2[(1.55^{0.5} - 1) + 1 - 0.77] = 0.716$$

$$E_3 = \frac{1}{8} + \frac{1}{36}[4+12] = 0.569$$

Since the effective utilization is less that the utilization bound, task τ_3 will meet its first deadline.

Test task τ_2:

$S_{2g} = \{\}$ $N_{2g} = 0$

$S_{2l} = \{\tau_1\}$ $N_{2l} = 1$

$$UB_2 = (1+1)\left(2^{\frac{1}{1+1}} - 1\right) = 0.828$$

$$E_2 = \frac{1}{8} + \frac{1}{60}[12] = 0.325$$

Since the effective utilization is less that the utilization bound, task τ_2 will meet its first deadline.

Test task τ_1:

$S_{1g} = \{\}$ $N_{1g} = 0$

$S_{1l} = \{\}$ $N_{1l} = 0$

$$UB_1 = \frac{d_1}{p_1} = \frac{2}{8} = 0.25$$

$$E_1 = \frac{1}{8} = 0.125$$

Since the effective utilization is less that the utilization bound, task τ_1 will meet its first deadline.

The results of the application of the utilization bound tests are summarized here:

- The given priorities will produce a valid schedule for tasks τ_1, τ_2, τ_3, and τ_4 since the effective utilization of each of these tasks does not exceed the corresponding utilization bound.

- In the case of task τ_5, the utilization bound test failed to show that the given priorities will produce a valid schedule. The test did not show that a valid schedule could not be produced.

The utilization bound tests constitute sufficient but not necessary conditions. They do not conclusively answer the question of the schedulability of the task system. Schedule construction or some other conclusive method is required for this purpose.

For situations in which the utilization bound test is inconclusive, a test that provides a conclusive answer must be supplied. But this needs to be applied only to those tasks for which the utilization bound test fails. The obvious conclusive test is to construct the schedule, but schedule construction is tedious. An analytical method exists as an alternative to the actual graphical construction of the schedule. It can be easily implemented as a software algorithm on a computer. The algorithm involves the solution of an iterative equation for the first completion times of the tasks in a task system. If the first completion time of a task does not exceed the associated deadline, then the task meets its first deadline and, consequently, meets all deadlines.

The iterative equation for the completion time of a task is given by

$$C_n(i) = e_n + b_n + \sum_{j=1}^{n-1} \left\lceil \frac{C_n(i-1)}{p_j} \right\rceil e_j \qquad \text{(EQ 5.3.7)}$$

where $C_n()$ is the completion time for the task computed iteratively until a stable value is attained, e_i is the task execution time, b_n is the worst-case task-blocking time, p_j is the task period, and the operator $\lceil \frac{a}{b} \rceil$ means the smallest integer greater than or equal to $\frac{a}{b}$. The initial value is given by

$$C_n(0) = b_n + \sum_{j=1}^{n} e_j \qquad \text{(EQ 5.3.8)}$$

The objective is to iteratively apply the above equation until one of two results is obtained.

- The iteration converges with a completion time less than the deadline of task τ_n, in which case, task τ_n will meet its first deadline—and, consequently, it will meet the deadlines for all release times.
- The iteration continues until the current value of the completion time exceeds the deadline of task τ_n. In. this case, task τ_n will miss its first deadline.

The above equation applies individually to all tasks in a task system. If all tasks meet their respective deadlines, then the task system is schedulable. That each task meet its deadline is, of course, a necessary condition for schedulability; the completion time test is a conclusive test for schedulability of a task system. In most situations, it is not necessary to apply the completion time test to all tasks in a task system.

Example 5.3.2: Calculating the precise completion times for periodic tasks.

In the previous Example 5.3.1 of a five-task system with a specified priority list, four of the tasks were shown by the utilization bound test to meet their first deadline. The fifth task, task τ_5, failed the test and, hence, no conclusion about its schedulability could be drawn. In lieu of actually constructing the schedule in order to determine the schedulability of task τ_5, the completion time test described above can be applied. Because tasks τ_1, τ_2, τ_3, and τ_4 have already been shown to meet their first deadline using the application of the utilization bound test, only task τ_5 need be considered.

For convenience, the specification of the task system of Example 5.3.1 is repeated below.

$N = number\ of\ tasks = 5$

$M = \text{number of processors} = 1$

Period	Deadline	Execution Time	Blocking Time
$p_1 = 8$	$d_1 = 2$	$e_1 = 1$	$b_1 = 0$
$p_2 = 60$	$d_2 = 60$	$e_2 = 16$	$b_2 = 0$
$p_3 = 36$	$d_3 = 28$	$e_3 = 4$	$b_3 = 0$
$p_4 = 50$	$d_4 = 30$	$e_4 = 2$	$b_4 = 1$
$p_5 = 30$	$d_5 = 30$	$e_5 = 2$	$b_5 = 0$

Priority list $= \{\tau_1, \tau_2, \tau_3, \tau_4, \tau_5\}$

Apply the completion time computation to task τ_5. For the initial value one obtains:

$$C_5(0) = 0 + \sum_{j=1}^{5} e_j = 1 + 16 + 4 + 2 + 2 = 25$$

Going through the iterations:

$i = 1$:

$$C_5(1) = e_5 + b_5 + \sum_{j=1}^{4} \left\lceil \frac{C_5(0)}{p_j} \right\rceil e_j = 2 + 0 + 4 + 16 + 4 + 2 = 28$$

$i = 2$:

$$C_5(2) = e_5 + b_5 + \sum_{j=1}^{4} \left\lceil \frac{C_5(1)}{p_j} \right\rceil e_j = 2 + 0 + 4 + 16 + 4 + 2 = 28$$

The iteration has converged and indicates that the first completion of task τ_5 will occur at time $t = 28$, which is less than the deadline value $d_5 = 30$.

The calculation of the completion time of a task by the above method is precise as no approximations or worst-case bounding values are involved. To use the completion time test to show that a task system is schedulable using a given priority list, the completion time of each task must be determined. An easier approach to the problem of determining the schedulability of a task system is to, first, apply the utilization bound test to all tasks, and to, then, apply the completion time method to those tasks for which the result of the utilization bound test is inconclusive. This is essentially the method applied to the example problem.

5.3.5 Assignment of Task Priorities

The methods of schedulability analysis described in the previous subsection presumed the existence of a task priority list. The highest priority tasks should be executing at any particular time assuming that they can be executed (not waiting on a resource). Tasks are preemptively assigned to processors in accordance with this priority list in a manner that ensures this rule. The order in which the tasks appear on the priority list is arbitrary and will be based on the system requirements specification. The analysis needs to be necessarily blind to the reasoning behind the priority list and for a given system; one task priority list can produce a valid schedule while another one may not.

For a given task system and M identical processors on which the task system is to be scheduled, there are two problems facing the designer.

1. The basic YES/NO question, is the task system schedulable on M processors?, needs an answer.
2. If the answer to the question in (1) is YES, then a valid schedule is desired.

The paradoxical aspect of this situation is that for the general task-scheduling problem, the only way to answer the question in (1) is to construct the schedule in (2). However, for some special cases the answer to the YES/NO question is easily obtained. As seen earlier, when all task deadlines are equal to periods, the processor utilization equation provides an easy answer. Other special cases exhibit polynomial time optimal algorithms for constructing the schedule. In such cases, the answer to the question in (1) above can be obtained by attempting to construct a valid schedule using the algorithm. It is important to recall that a scheduling algorithm is said to be *optimal* if it always produces a *valid* schedule for every task system that is *feasible*.

By definition, if the task system is feasible, an *optimal* algorithm will produce a valid schedule. The known optimal task scheduling algorithms are all priority list algorithms. Such an algorithm defines the basis upon which a priority list of the tasks is constructed. The tasks are, then, scheduled preemptively in a manner that ensures the highest priority tasks are executing at any particular time. The *intrinsic importance* of a task (some measure of how vital the service performed by the task is to the overall functioning of the real time system) has no bearing on the assigned priority of the task. It should be noted that intrinsic importance is a subjective measure, and the schedulability analysis algorithms necessarily require some general objective measure that can actually be calculated either during schedule creation or while the real time system is in operation.

With this kind of priority assignment, it may well happen that a task of relatively low intrinsic importance is assigned a higher priority than a task of relatively high intrinsic importance. Task priorities are assigned in a manner that will ensure that *all* tasks meet their deadlines even if this means relatively unimportant (intrinsically) tasks are assigned high priorities. This is especially important for hard real time systems where deadlines cannot be missed.

Three important task priority assignment algorithms from the real time systems field will be introduced with examples. They include two algorithms that use static priorities and one algorithm that uses dynamic priorities.

- The *deadline-monotonic algorithm* assigns task priorities in order of increasing deadlines. This is a *static* priority assignment algorithm.
- The *rate-monotonic algorithm* assigns task priorities in order of increasing periods, i.e., decreasing rates (frequencies). This is a *static* priority assignment algorithm.
- The *earliest deadline algorithm* schedules tasks in the following manner: at each instant of time, task priorities are assigned in increasing order of currently impending deadlines. This is a *dynamic* priority assignment algorithm since the priority assignments vary as the execution proceeds.

Each of these algorithms is optimal for a certain class of task-scheduling problem, and each will be examined in detail in the following sections.

5.4 Deadline-Monotonic Scheduling

The deadline-monotonic scheduling algorithm assigns priorities to tasks in the order of *increasing deadlines* d_n. The priorities are static and do not change once they are assigned since a task deadline is a fixed parameter of the task system, determined *a priori* by the system engineers. First, for the static scheduling case—the designer applies the algorithm to create a task priority list and then creates a schedule by preemptively allocating tasks to processors based on their priority. Second, for the dynamic scheduling case, the designer again applies the algorithm to create a task priority list and then creates task control blocks or instances in the software based on these priorities (in a modern RTOS, task priority is a parameter supplied as a part of the system call when the task is created). It can then be used to predict the behavior of the task system when executed by the multitasking executive (RTOS).

The deadline-monotonic priority assignment algorithm has the very useful property in that it is an optimal static priority assignment algorithm for *synchronous task systems* executing on *one processor*. Recalling earlier definitions—an optimal task-scheduling algorithm is one that will result in a valid task schedule if the task system is schedulable (even though the solution is not guaranteed to be unique). A task system is said to be schedulable on M identical processors *if there is a valid schedule produced by a static task priority assignment* (as opposed to just feasible). For a single processor case, $M = 1$ and a pseudo-polynomial time algorithm can be formulated to decide if the system is schedulable (remember the definition of schedulable from the previous line). The logic is as follows.

- As determined earlier for synchronous task systems on one processor, the pseudo-polynomial time algorithm to check for validity of a schedule for a given priority assignment consists of the construction of the schedule from a critical instant through the longest period.
- Based on the earlier definition of algorithm optimality, the deadline-monotonic priority assignment being optimal (for synchronous task systems on one processor only) means it will produce a valid schedule if the task system is feasible (note that feasible means a valid schedule can be produced and does not have the requirement of static priorities that schedulable does).

- Therefore, one merely needs to assign task priorities in accordance with the deadline-monotonic algorithm and, then, construct the schedule through the longest period. If all tasks meet their first deadlines, the task system is feasible. Otherwise, it is not.

In place of schedule construction, analytical methods can be used.

Example 5.4.1: The use of deadline monotonic scheduling.

The use of deadline-monotonic scheduling will be illustrated by the construction of a schedule for the task system shown below.

$N = number\ of\ tasks = 3$

Period	Deadline	Execution Time
$p_1 = 4$	$d_1 = 2$	$e_1 = 1$
$p_2 = 6$	$d_2 = 4$	$e_2 = 2$
$p_3 = 10$	$d_3 = 10$	$e_3 = 3$

The processor utilization is determined to be $U = \frac{1}{4} + \frac{2}{6} + \frac{3}{10} = 0.883 < 1$.

The deadline utilization is determined to be $D = \frac{1}{2} + \frac{2}{4} + \frac{3}{10} = 1.3 > 1$.

The first result—since the test of Equation 5.3.2 is necessary but not sufficient—implies that this test cannot show that the task system cannot be scheduled. The second result—since the test of Equation 5.3.3 is sufficient but not necessary—implies that this test cannot show that the task system can be scheduled. The only known algorithm for answering the YES/NO question, is this task system schedulable?, is to construct the schedule through the longest period using an optimal static priority assignment. If all tasks meet their first deadline, they will meet their deadlines for all task releases, and the task system is schedulable. As an alternative to graphical schedule construction, analytical techniques can be applied since application of these analytical techniques is the equivalent of schedule construction.

The result of schedule construction using the deadline-monotonic priority assignment is shown in Figure 5.18. The deadline-monotonic priority assignment is an optimal static priority assignment in this case. Each indication of a release time is accompanied by the associated deadline. Since the tasks are arbitrarily preemptable, the algorithm is free to suspend an executing task at any point so that a higher priority task can execute. That is precisely what happens to task τ_3 at times 4 and 6. The schedule shows that all tasks meet their first deadlines, thus, proving that the task system is schedulable.

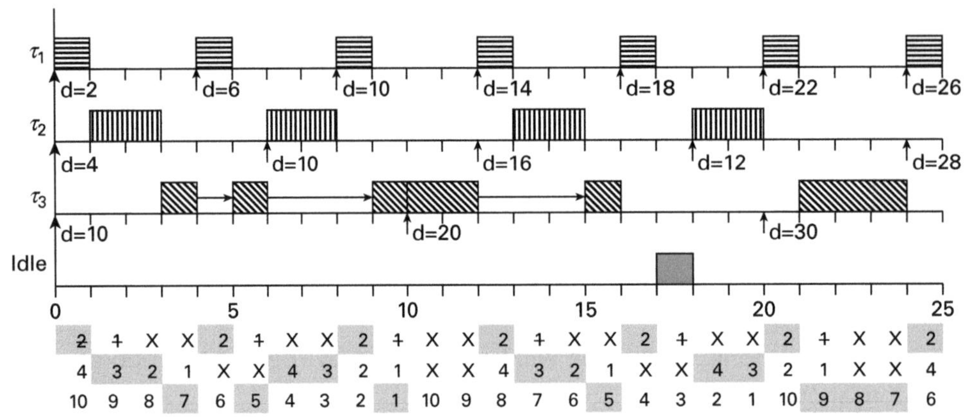

Figure 5.18 A Deadline-Monotonic Schedule.

The concept of optimality of scheduling algorithms applied in this situation merely states that if a synchronous task system is schedulable on one processor, the deadline-monotonic algorithm will produce a valid schedule. The schedule need not be and seldom is unique. There are, for example, several easily identified ways in which the schedule of Figure 5.18 can be modified and remain valid.

Example 5.4.2: Application of the completion time method.

It is informative to apply the completion time calculation to this task-scheduling problem. The above iterative equation must be applied:

Calculate the completion time for task τ_3:

$$C_3(0) = 1+2+3 = 6$$

$$C_3(1) = 3 + \left\lceil \frac{6}{6} \right\rceil 2 + \left\lceil \frac{6}{4} \right\rceil 1 = 7$$

$$C_3(2) = 3 + \left\lceil \frac{7}{6} \right\rceil 2 + \left\lceil \frac{7}{4} \right\rceil 1 = 9$$

$$C_3(3) = 3 + \left\lceil \frac{9}{6} \right\rceil 2 + \left\lceil \frac{9}{4} \right\rceil 1 = 10$$

$$C_3(4) = 3 + \left\lceil \frac{10}{6} \right\rceil 2 + \left\lceil \frac{10}{4} \right\rceil 1 = 10$$

The iteration has converged, indicating that task τ_3 will complete at time $t = 10$, and this is verified by examination of the task schedule of Figure 5.18.

Calculate the completion time of task τ_2:

$$C_2(0) = 1+2 = 3$$

$$C_2(1) = 2 + \left\lceil \frac{2}{4} \right\rceil 1 = 3$$

The iteration has converged, indicating that task τ_2 will complete at time $t = 3$, and this is verified by examination of the task schedule of Figure 5.18.

This example demonstrates the equivalency of schedule construction and the analytical solution for completion times.

A summary of deadline-monotonic task scheduling is presented below.

- The deadline-monotonic algorithm requires a task set that is preemptable and independent, and it uses static task priorities.
- The deadline-monotonic algorithm is optimal for synchronous task systems on one processor.
- There is no simple condition that is both necessary and sufficient for determining if a task system is schedulable by deadline-monotonic scheduling, but a utilization bound can be obtained using the equations in the previous section.
- A synchronous task system on one processor can be tested by constructing the schedule through the longest task period, and this can be accomplished in pseudo-polynomial time. Either graphical or analytical methods can be used.

5.5 Rate-Monotonic Scheduling

The rate-monotonic priority assignment algorithm is a static priority assignment algorithm that assigns task priorities in the order of increasing periods. For task systems in which deadlines equal periods ($d_n = p_n$ for all tasks τ_n), the rate-monotonic and deadline-monotonic algorithms are equivalent, which results in the following conclusion.

For task systems in which $d_n = p_n$ for all tasks τ_n, rate-monotonic priority assignment is an optimal static assignment algorithm for asynchronous task systems on one processor.

Using the concept of a critical instant, a worst-case performance bound for rate monotonic scheduling can be developed. This result follows directly from the developments of Section 5.3 under the condition that $d_n = p_n$ for all tasks τ_n for, in this case, the following simplifications occur.

The set of tasks S_{ng} consisting of those tasks with priorities *greater than or equal to* that of task τ_n and with periods *greater than or equal to* the deadline of task τ_n is empty. That is, $S_{ng} = \{\}$, and $N_{ng} = 0$. The set of tasks S_{nl} consisting of those tasks with priorities *higher* than that of task τ_n but with *periods less than* the deadline of task τ_n contains all tasks with priorities greater than that of task τ_n. That is, $S_{nl} = \{\tau_1, \tau_2, \cdots, \tau_{n-1}\}$ and $N_{nl} = n-1$.

The effective utilization of task τ_n, then, reduces to

$$E_n = \sum_{i=1}^{N} \frac{e_i}{p_i} + \frac{b_n}{p_n} \tag{EQ 5.5.1}$$

and the utilization bound becomes simply

$$E_n = n\left(2^{\frac{1}{n}} - 1\right) \tag{EQ 5.5.2}$$

Furthermore, in the situation in which there is no blocking ($b_n = 0$), if task τ_n meets its deadline, then so also do all higher priority tasks. This conclusion is evident from the behavior of Equation 5.5.1, with $b_n = 0$ and equation 5.5.2 when n is replaced by $n - 1$.

The rate-monotonic scheduling algorithm will produce a valid schedule for a synchronous system of N independent tasks if—but only if—the following inequality holds.

$$U \leq N\left(2^{\frac{1}{N}} - 1\right) \qquad \text{(EQ 5.5.3)}$$

In the limit, as N grows large, the right hand side of this inequality approaches $ln\ 2 = 0.692$, implying that processor utilization can be limited to values as low as 69% solely by the scheduling process. The inequality of Equation 5.5.3 is a worst-case bound and, hence, can serve to determine that a task system is schedulable. It cannot be used to determine if a task system is *not* schedulable.

Example 5.5.1: A simple example of rate-monotonic task scheduling.

A simple task system consisting of three periodic tasks is specified below.

$N = number\ of\ tasks = 3$

Period	Execution Time
$p_1 = 20$	$e_1 = 4$
$p_2 = 30$	$e_2 = 8$
$p_3 = 70$	$e_3 = 20$

The processor utilization is determined to be $U = \frac{4}{20} + \frac{8}{30} + \frac{20}{70} = 0.752.$

The utilization bound is $3\left(2^{\frac{1}{3}} - 1\right) = 0.780.$

Since the processor utilization is less than the utilization bound, this task system is schedulable using rate-monotonic priority assignment. This conclusion is verified by the rate monotonic schedule depicted in Figure 5.19, in which the low priority task τ_3 is preempted twice but still completes prior to its deadline at $t = 70$.

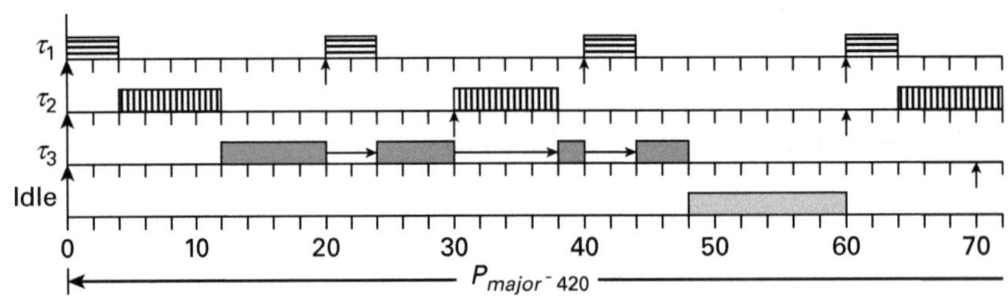

Figure 5.19 A Rate-Monotonic Task Schedule with Processor Utilization = 0.752.

If the execution time of task τ_1 is increased from its value of 4 to a value of 8, the processor utilization increases to a value $U=\frac{8}{20}+\frac{8}{30}+\frac{20}{70}=0.952$. Since this value is greater than that of the utilization bound, no conclusion can be drawn concerning the schedulability of the task system. However, since the rate-monotonic algorithm is optimal, in this particular case, the schedulability question can be answered by constructing the schedules using the rate-monotonic priority assignment. The resulting schedule is shown in Figure 5.20 and illustrates that all tasks meet their first deadlines and, hence, that the task system is schedulable. Note that the low priority task τ_3 is preempted three times but, yet, completes prior to its deadline at $t = 70$.

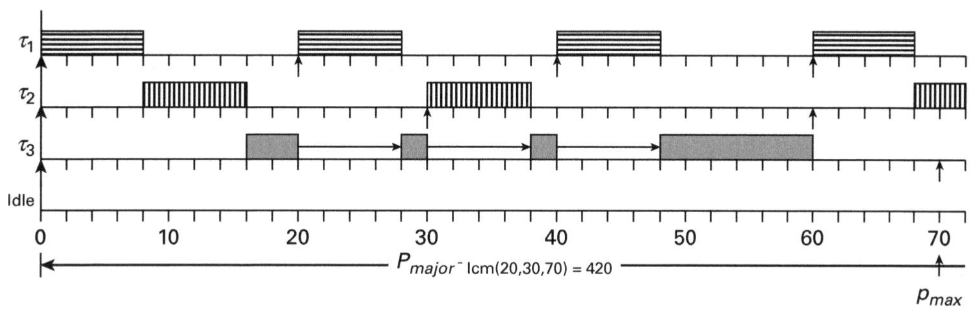

Figure 5.20 A Rate-Monotonic Schedule with Processor Utilization = 0.952.

It should be noted that the schedule of Figure 5.20 shows no processor idle time, yet the processor utilization was computed to be 0.952. This apparent conflict between the analytical and schedule constructing methods can be explained. The cyclic schedules produced by these algorithms repeat at the major cycle period. Yet, the schedule construction or, equivalently, the analytical method concern only the first release time of each task. That is, the schedule need be constructed only through the longest period. The reason it is not necessary to examine the entire major cycle is that for synchronous task systems, if a task meets its first deadline, it will meet all deadlines for all releases. It is entirely possible for the processor utilization over the first duration of the longest period to be different from that over a complete major cycle.

Example 5.5.2: A second example of rate monotonic scheduling.

A second example of rate-monotonic scheduling—and one that illustrates the phenomenon described above—will be examined. In the treatment of this problem, the schedule over the longest period will be compared to the schedule as it exists over the complete major cycle. Since the pattern of task execution is periodic at the major cycle rate, the task utilization over the major cycle determines the long-term utilization. The task system to be considered is described below.

$N = number\ of\ tasks = 3$

Period	Execution Time
$p_1 = 4$	$e_1 = 1$
$p_2 = 6$	$e_2 = 2$
$p_3 = 10$	$e_3 = 3$

The processor utilization is determined to be $U = \frac{1}{4} + \frac{2}{6} + \frac{3}{10} = 0.883$.

The utilization bound for three tasks using rate-monotonic scheduling is 0.780, making the utilization bound test inconclusive. Schedule construction over the longest period yields the result depicted in Figure 5.21 and indicates that the task system is schedulable. There is no processor idle time indicated in this schedule even though the processor utilization was calculated to be 0.883. The resolution of this apparent discrepancy can be obtained by examining the task schedule over the complete major cycle, which, in this case, has a duration of 60. A complete major cycle is depicted in Figure 5.22 and shows processor idle time appearing over the remainder of the major cycle. The total processor idle time over the entire major cycle represents 11.7%, as required.

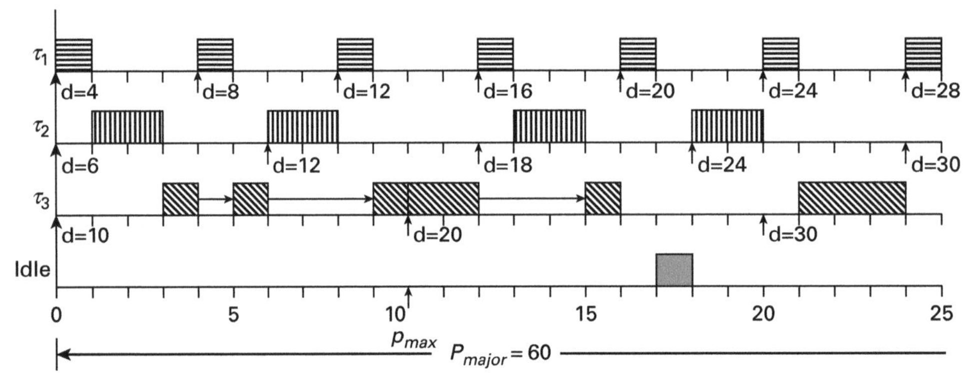

Figure 5.21 A Rate-Monotonic Schedule with Processor Utilization = 0.883.

A summary of rate-monotonic task scheduling is presented below.

- The rate-monotonic algorithm requires a task set that is preemptable and independent and uses static task priorities.
- Under the condition that deadlines equal periods, the rate-monotonic priority assignment is an optimal static priority assignment for *synchronous task systems on one processor.*
- There is no simple condition that is both necessary and sufficient for determining if a task system is schedulable by rate-monotonic scheduling, but a utilization bound can be simply stated as in Equation 5.5.3.
- A *synchronous task system on one processor* can be tested by constructing the schedule through the longest task period, and this can be accomplished in pseudopolynomial time. Alternatively, analytic completion time algorithms can be used.

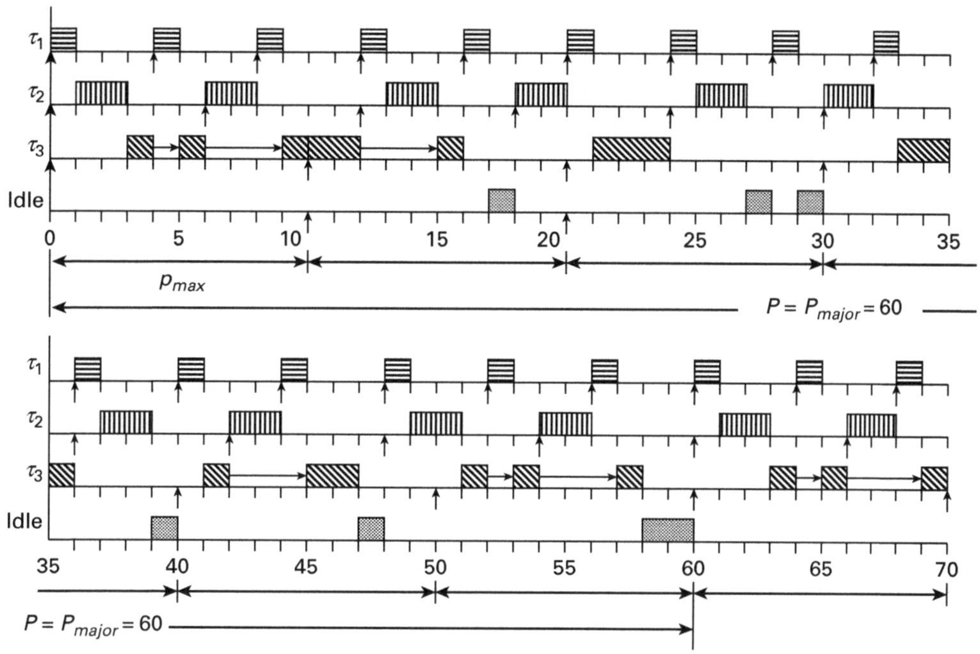

Figure 5.22 A Complete Major Cycle.

5.6 Earliest-Deadline Scheduling

The deadline-monotonic and rate-monotonic algorithms are static priority assignment algorithms. This means that the task priorities do not change during system operation when the task system is executing under control of a multitasking executive (RTOS). In the case of a cyclic executive, a static schedule is constructed in advance for the system, and the task priorities do not change as the schedule is being developed. The earliest deadline algorithm, on the other hand, is a dynamic priority assignment algorithm, which means that the assigned priorities can change during execution by an RTOS or during the development of a static cyclic schedule. The earliest deadline priority assignment is made as follows.

At each instant of time, task priorities are assigned in increasing order of currently impending deadlines.

The analogy to the Critical Path Scheduling algorithm from the aperiodic scheduling section is evident. Application of the algorithm requires a continuous determination of the *"time-to-deadline"* for each task in the task system, and task priorities are continuously assigned according to these times. The term *continuous* is clear when the algorithm is applied to the determination of a static schedule by graphical or analytical means. What does it mean for a real time operating system? The execution of the tasks is essentially in the form of a computer program. This is inherently a discrete time process controlled by the scheduler, which typically takes these decisions (after recomputing task priorities) upon each *tick of the real time clock*. The earliest deadline algorithm has optimal properties similar to the prior algorithms (always produces a *valid* schedule for every task system that is *feasible)*.

The earliest deadline algorithm is an optimal dynamic priority assignment algorithm for scheduling periodic tasks on a single processor.

That the earliest deadline algorithm is optimal is particularly significant in view of the fact that Equation 5.3.2 represents a *necessary and sufficient* condition for a synchronous task system to be feasible (if there is a *valid* schedule for the task system) when deadlines equal periods, and this leads to the following result.

For synchronous task systems in which deadlines equal periods, a necessary and sufficient condition for a task system to be feasible is $U \le 1$. If these conditions are met, earliest deadline scheduling will produce a valid schedule.

The term *feasible* is extremely important, implying no requirement for static priorities as opposed to the term *schedulable* (which requires a schedule based on static priorities). Further, for *arbitrary task systems* that are not synchronous and also with task deadlines that differ from task periods, the earliest deadline algorithm is still optimal with respect to dynamic priority assignment on a single processor.

The following example illustrates the application of the earliest deadline scheduling algorithm. For simplicity, the example task system has $deadlines = periods$, but the method is applicable, as well, when this condition does not hold.

Example 5.6.1: An application of earliest deadline scheduling.

The application of earliest deadline scheduling to the task system specified below will be used to determine if the task system is feasible.

$N = number\ of\ tasks = 3$

Period	Deadline	Execution Time
$p_1 = 4$	$d_1 = 4$	$e_1 = 1$
$p_2 = 6$	$d_2 = 6$	$e_2 = 2$
$p_3 = 8$	$d_3 = 8$	$e_3 = 3$

The processor utilization is determined to be $U = \frac{1}{4} + \frac{2}{6} + \frac{3}{8} = 0.958 < 1.$

The rate-monotonic utilization bound for three tasks is 0.780; the test is inconclusive. The only manner in which the YES/NO question, is this task system feasible?, can be answered is to construct a schedule showing that all tasks meet their first deadline. Since the earliest deadline algorithm is an optimal dynamic priority assignment, application of this algorithm will produce a valid schedule if the task system is feasible.

The earliest deadline schedule is depicted in Figure 5.23 and indicates that all tasks meet their first deadline. This task system is, therefore, *feasible* on a single processor; it can be scheduled on a single processor but is not schedulable on a single processor. That is, there is no *static* priority assignment that will produce a valid schedule. That the task system is not schedulable may be verified by attempting to apply the rate-monotonic algorithm—it will not result in a valid schedule. Since the rate-monotonic algorithm is an optimal static priority assignment algorithm, if it does not produce a valid schedule, neither will any other static priority algorithm.

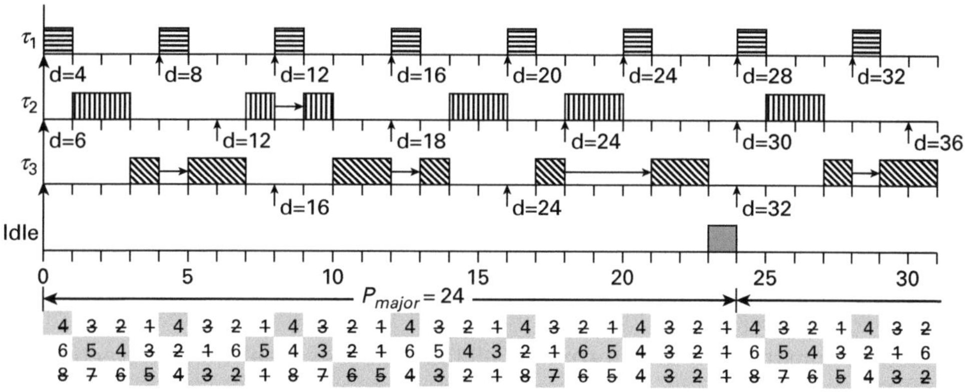

Figure 5.23 An Earliest Deadline Task Schedule.

The schedulability analysis of the previous sections deals only with synchronous task systems. In synchronous task systems, all tasks are released in phase at time $t = 0$. This permits the easy identification of a critical instance. For *asynchronous* task systems, the complexity of the task-scheduling problem increases considerably. *There is no optimal static priority assignment* as shown in the case of synchronous task systems. In particular, it should be noted that the deadline-monotonic algorithm is not optimal for asynchronous task systems for *any* number of processors.

5.7 Multiprocessor Scheduling

When periodic tasks are to be scheduled on more than one processor, the utilization bound equations apply and are repeated here as Equations 5.7.1 and 5.7.2. Equation 5.7.1 is a *necessary but not sufficient* condition while Equation 5.7.2 is a *sufficient but not necessary* condition. When deadlines equal periods, these inequalities are equivalent and represent a necessary and sufficient condition for feasibility.

$$U = \sum_{1}^{N} \frac{e_n}{p_n} \leq M \qquad \text{(EQ 5.7.1)}$$

$$D = \sum_{1}^{N} \frac{e_n}{d_n} \leq M \qquad \text{(EQ 5.7.2)}$$

If the feasibility of a task system is not evident from the utilization bound of Equation 5.7.2, the only known manner in which feasibility can be shown is construction of a valid schedule.

Even if a task system is known to be feasible on M (greater than one) processors, *there is no polynomial time algorithm guaranteed to produce a valid schedule*. That is, there is no known optimal scheduling algorithm. The usual approach to schedule construction is the use of some systematic technique as a heuristic method, such as rate-monotonic scheduling, deadline-monotonic scheduling, earliest deadline scheduling, or some other technique.

When scheduling tasks on multiple processors, there are two possible methods for determining which tasks will be assigned to which processors. Determining whether a task system (either synchronous or asynchronous) is *schedulable* (the schedulable term implies static priority assignment) on multiple processors using either of these methods is, in general, a very difficult problem.

- In the *non-partitioning method,* the M processors are treated as a single resource, and the M tasks with the highest priorities are always executing, with the priorities assigned by whatever heuristic method has been chosen.
- In the *partitioning method*, tasks are partitioned into separate groups, possibly in some optimal or near optimal manner, and each group is then scheduled on a single processor. Tasks are assigned static priorities within a group using whatever heuristic method has been chosen.

Once again, when the algorithm for determining the schedulability or feasibility of a task system consists of schedule construction to determine if all task releases meet their respective deadlines, the question of just how much of the infinite periodic schedule need be constructed to prove feasibility must be addressed. The answer to this question is given below for task systems with *static* priorities.

- For *synchronous* task systems, the schedule is periodic with period $P = lcm(p_1, p_2, ..., p_N)$ so that the schedule must be constructed through time $t = P$.
- For *asynchronous* task systems, the schedule becomes periodic at time $r_{max} + P$ where r_{max} is equal to the largest of the initial release times of all the tasks in the system. Thus, it is necessary to construct the schedule through time $t = r_{max} + 2P$.

For synchronous task systems on one processor, it is necessary to construct the schedule only through the longest period. The additional complexity of multiprocessor scheduling requires schedule construction over the longer base period P. In either case, the problem remains generally very difficult, implying that testing a static priority assignment to determine if a valid schedule is produced can be difficult.

5.7.1 Partitioning and Non-Partitioning Methods

An immediate question that arises concerning the partitioning and non-partitioning methods of heuristically scheduling periodic tasks on multiple processors concerns the relative effectiveness of the two methods. There appears to be no clear-cut answer to this question in the sense that which method is most effective will depend on the particular application. This circumstance can be illustrated by the following two simple examples.

Example 5.7.1: Scheduling tasks on two processors by the partitioning and non-partitioning methods.

The first example comparing the partitioning and non-partitioning methods involves the following task system.

$N = number\ of\ tasks = 4$

$\mathrm{M = number\ of\ processors} = 2$

Period	Deadline	Execution Time
$p_1 = 2$	$d_1 = 2$	$e_1 = 1$
$p_2 = 3$	$d_2 = 3$	$e_2 = 2$
$p_3 = 4$	$d_3 = 4$	$e_3 = 2$
$p_4 = 6$	$d_4 = 6$	$e_4 = 2$

The processor utilization is determined to be

$$U = \frac{1}{2} + \frac{2}{3} + \frac{2}{4} + \frac{2}{6} = 2.0 \leq 2.$$

Even though the task system is known to be feasible, on the basis that deadlines equal periods and that it satisfies this utilization bound, there is no known algorithm guaranteed to produce a valid schedule. Heuristic algorithms can be applied but with no guarantee of success. If deadline-monotonic scheduling is selected as the heuristic algorithm, then either a partitioning or non-partitioning approach must be used.

The result of using a partitioning approach with deadline-monotonic priority assignments is depicted in the task schedule contained in Figure 5.24. In constructing this schedule, the following partitioning was used.

Tasks τ_1 and τ_3 are assigned to processor P_1 for execution.

Tasks τ_2 and τ_4 are assigned to processor P_2 for execution.

That is, tasks τ_1 and τ_3 are scheduled on processor P_1, using deadline-monotonic scheduling (since deadlines equal periods in this example, deadline-monotonic and rate-monotonic priority assignments are equivalent) independently of the scheduling of tasks τ_2 and τ_4 on processor P_2, which is also accomplished using deadline-monotonic scheduling. The general requirement of a valid schedule for a synchronous task system on multiple processors is that all tasks meet their deadlines in the interval $P = lcm(2, 3, 4, 6) = 12$, the base period of the task system.

In this case, however, because of the manner of partitioning into two task systems that are independently scheduled on the two processors, the single processor requirement for schedule validity can be used. Since the task systems are synchronous and scheduled on a single processor, the schedule need only be constructed through the longest period—time $t = 4$ for processor P_1 and time $t = 6$ for processor P_2. All tasks meet their first deadlines, and, hence, the schedule is valid.

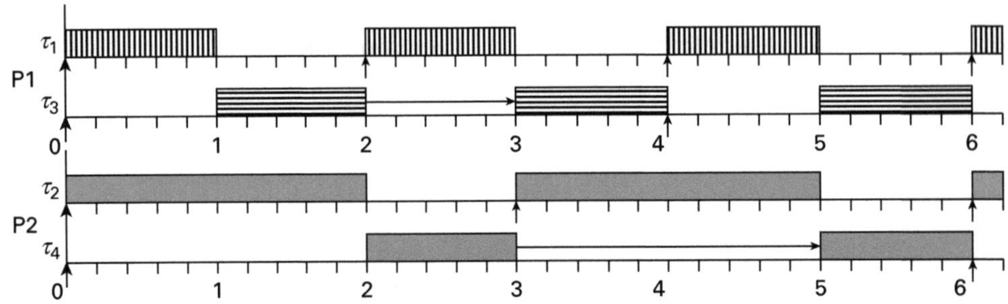

Figure 5.24 The Partitioning Method Applied to Example 5.7.1.

An attempt to schedule the same task system using the non-partitioning method with deadline-monotonic scheduling is shown in Figure 5.25. The release times of the tasks are indicated along the lower bar in the diagram of Figure 5.25. In this case, the combination of the two processors is considered a common pool of computational resources which are assigned to execute the tasks according to the deadline-monotonic priority assignment. That is, the two highest priority tasks are executing at any particular time, without regard to the particular processor on which they are scheduled. In this case, the general requirement of a valid schedule—that all tasks meet their deadlines in the interval $P = lcm(2, 3, 4, 6) = 12$ —the base period of the task system must be applied.

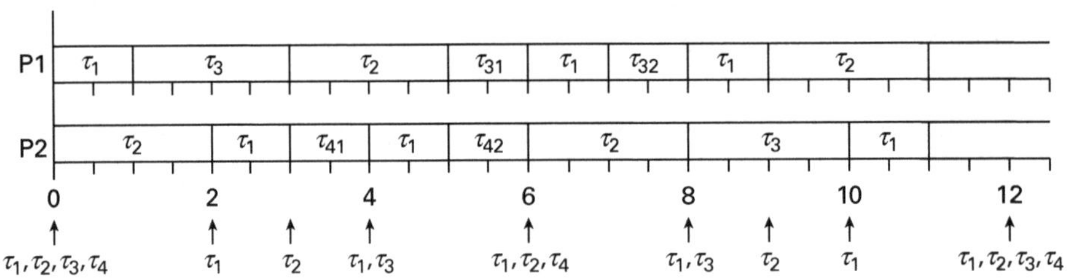

Figure 5.25 The Non-Partitioning Method Applied to Example 5.7.1.

As can be seen from the diagram of Figure 5.25, the deadline-monotonic priority assignment fails to produce a valid schedule. At time $t = 6$, task τ_4 is released, but higher priority tasks τ_1, τ_2, and τ_3 consume all execution time until time $t = 11$, at which point, there is not enough time on either processor for τ_4 to complete by its deadline at time $t = 12$. (Note: Tasks may be preempted and split into multiple parts which is denoted by an additional subscript. τ_{41}, and τ_{42} denote that the task was preempted and split into two parts for execution.)

In this example, then, the partitioning method, together with deadline-monotonic scheduling used as a heuristic algorithm, produces a valid schedule while the non-partitioning method used with the same heuristic algorithm fails to produce a valid schedule.

Example 5.7.2: A second example of scheduling tasks on two processors.

The second example of the scheduling of tasks on multiple processors using both the partitioning and non-partitioning methods involves the following task system.

$N = number\ of\ tasks = 4$

M = number of processors = 2

Period	Deadline	Execution Time
$p_1 = 20$	$d_1 = 20$	$e_1 = 10$
$p_2 = 30$	$d_2 = 30$	$e_2 = 11$
$p_3 = 30$	$d_3 = 30$	$e_3 = 21$
$p_4 = 40$	$d_4 = 40$	$e_4 = 8$

The processor utilization is determined to be

$$U = \frac{10}{20} + \frac{11}{30} + \frac{21}{30} + \frac{8}{40} = 1.767 \le 2.$$

Since deadlines = periods, this result is a necessary and sufficient condition for feasibility of the task system on two processors. Again, even though the task system is known to be feasible on the basis of the utilization bound test, there is no known optimal scheduling algorithm for producing a valid schedule. In this case, the deadline-monotonic scheduling algorithm is, again, applied as a heuristic algorithm, employing both the partitioning and non-partitioning methods. First, the non-partitioning method is applied, with the following partitioning.

Tasks τ_1 and τ_2 are assigned to processor P_1 for execution.

Tasks τ_3 and τ_4 are assigned to processor P_2 for execution.

The individual processor utilizations are

$$U_1 = \frac{10}{20} + \frac{11}{30} = 0.867 \le 1.$$

$$U_2 = \frac{21}{30} + \frac{8}{40} = 0.9 \le 1.$$

The result is the attempted schedule depicted in Figure 5.26, which fails when task τ_2 misses its first deadline at time t = 30. There is no other partitioning in which either U_1 or U_2—the processor utilization factors of tasks τ_1 and τ_2—do not exceed the value one, so the partitioning method clearly fails to produce a valid schedule for this task system.

When deadline-monotonic scheduling is applied to this task system using the non-partitioning method, the fact that two tasks, tasks τ_2 and τ_3, have the same period produces a tie for the second priority. The deadline-monotonic scheduling algorithm allows such ties to be arbitrarily broken, and, in this case the tie is broken to produce the priority list (τ_1, τ_2, τ_3, τ_4). The resulting schedule attempt is shown in the diagram of Figure 5.27, which shows task τ_3 missing its first

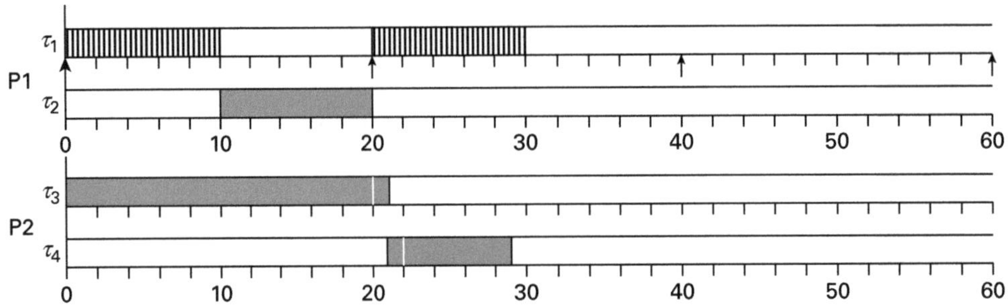

Figure 5.26 The Partitioning Method Applied to Example 5.7.2.

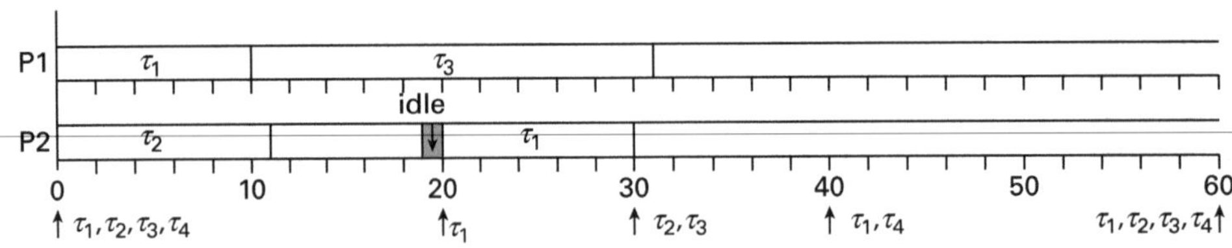

Figure 5.27 The Non-Partitioning Method Applied to Example 5.7.2 with Priorities $(\tau_1, \tau_2, \tau_3, \tau_4)$.

deadline at time $t = 30$. It would appear that deadline-monotonic scheduling used as a heuristic algorithm fails in this case.

However, if the tie encountered in forming the deadline-monotonic priority list is broken the other way and the priority list is changed to $(\tau_1, \tau_3, \tau_2, \tau_4)$, the valid schedule shown in Figure 5.28 results. Both priority lists represent the result of deadline-monotonic priority assignment, and the fact that one priority list produces a valid schedule while the other does not merely points out that the deadline-monotonic priority assignment is not optimal in this situation.

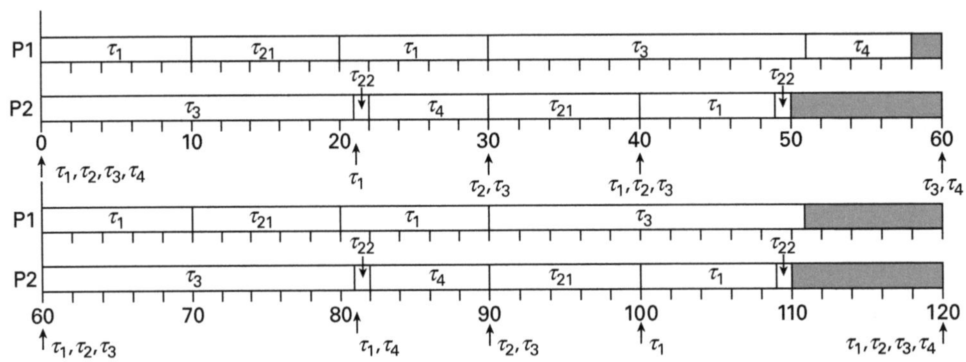

Figure 5.28 The Non-Partitioning Method Applied to Example 5.7.2 with Priorities $(\tau_1, \tau_3, \tau_2, \tau_4)$.

That is, since an optimal algorithm will always produce a valid schedule if one exists, and since both priority lists (even though they are different) represent deadline-monotonic priorities, if the deadline-monotonic algorithm were optimal, both priority lists would have produced valid schedules.

The two examples above deal with the heuristic use of the deadline-monotonic priority assignment algorithm, which is an algorithm for the assignment of *static* priorities. Might not a dynamic priority assignment algorithm demonstrate optimal performance? The answer to this question is no, and a simple example serves to illustrate the failure of one dynamic priority assignment algorithm, namely, the earliest deadline priority assignment to produce a valid schedule for a task system that is feasible on two processors.

Example 5.7.3: Application of the earliest deadline method to the scheduling of tasks on two processors.

The algorithm is applied in a manner that treats the multiple processors as a common pool of computational resources, i.e., the non-partitioning method is used. The task system for this example is specified below.

$N = number\ of\ tasks = 3$

M = number of processors = 2

Period	Deadline	Execution Time
$p_1 = 40$	$d_1 = 40$	$e_1 = 20$
$p_2 = 40$	$d_2 = 40$	$e_2 = 20$
$p_3 = 44$	$d_3 = 44$	$e_3 = 40$

The processor utilization is determined to be

$$U = \frac{20}{40} + \frac{20}{40} + \frac{40}{44} = 1.909 \leq 2 .$$

Since deadlines = periods, this condition is both necessary and sufficient, and, hence, the task system is feasible. In an attempt to find a valid schedule, the earliest deadline priority assignment algorithm will be used. The result is shown in Figure 5.29 where it can be seen that task τ_3 misses its first deadline. Thus, the earliest deadline priority assignment algorithm fails to produce a valid schedule even though the task system is feasible.

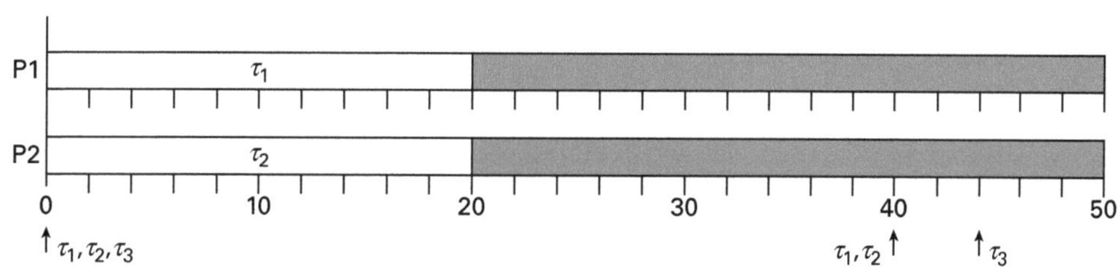

Figure 5.29 Earliest Deadline Scheduling.

A valid schedule for this task set is easily obtainable by assigning task τ_3 to processor P_1 and tasks τ_1 and τ_2 to processor P_2, as illustrated in the task schedule shown in Figure 5.30. If the earliest deadline priority assignment algorithm were optimal, it would have produced a valid schedule.

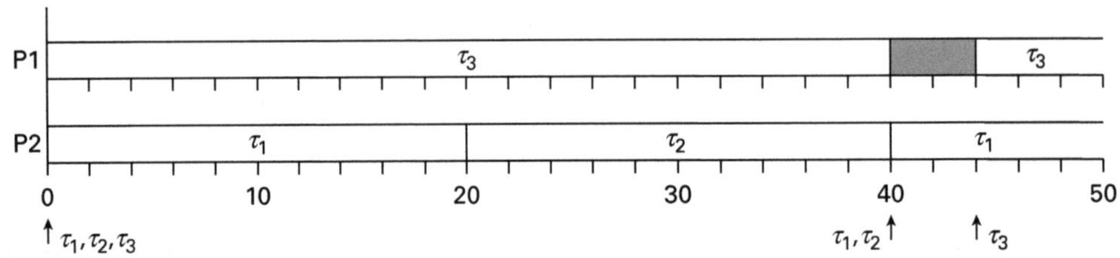

Figure 5.30 A Valid Schedule.

In summary, the following observations can be made concerning the scheduling of periodic tasks on multiple processors.

- Even if a task system is known to be feasible on M processors, there is no known algorithm guaranteed to produce a valid schedule, i.e., there is no known optimal algorithm.
- In practice, heuristic algorithms are used, and these algorithms are applied using either the partitioning or non-partitioning methods.
- The non-partitioning method suffers from the possibility of having time available simultaneously on more than one processor but not being able to use the totality of this time to satisfy the processing requirements of a single task. This was the difficulty in Example 5.7.1.
- The partitioning method suffers from the inflexibility of not being able to shift processing requirements from a busy partition to idle processors serving other partitions. This was the difficulty in Example 5.7.2.
- There is no definitive way in which the performance of the partitioning and non-partitioning methods can be compared.

5.7.2 Least Slack Time Priority Assignment

Before leaving the discussion of periodic task scheduling on multiple processors, another dynamic priority assignment algorithm will be described: the *least slack time* priority assignment algorithm. This algorithm requires that task priorities be assigned in the following manner.

At any instant of time, tasks are assigned priorities in increasing order of slack times. The slack time of a task τ_n at time t is defined as

$$t_s = t_d - e(t) - t \qquad \text{(EQ 5.7.3)}$$

where t_d is the time of the impending deadline of task τ_n, and $e(t)$ is execution time remaining in order to complete task τ_n.

The slack time t_d is, thus, the difference between the time available to meet the impending deadline and the execution time required to complete the task. The least slack time algorithm assigns tasks to processors using the concept of *processor sharing* introduced earlier where it was applied to the scheduling of aperiodic tasks on multiple processors in the presence of precedence constraints. Processor sharing schedules, it will be recalled, are not themselves realizable schedules, but they can be translated into realizable preemptive schedules.

The least slack time algorithm is not an optimal algorithm for scheduling periodic tasks on multiple processors (remember, there is no such polynomial time scheduling algorithm), but when used as a heuristic algorithm, and in the case of periodic task scheduling, on one processor, it has some desirable features.

- For any number of processors M, and for any task system, if the earliest deadline algorithm produces a valid schedule, so also does the least slack time algorithm.
- In the single processor case, the least slack time algorithm is optimal. This conclusion follows directly from the statement above. On a single processor, processor sharing is equivalent to arbitrary breaking of ties among task priorities.
- For multiple processors, there are task systems for which the least slack time algorithm will produce a valid schedule when the earliest deadline algorithm will not.
- The length of schedule construction required to determine if a schedule produced by the least slack time algorithm is valid is identical to that required of the earliest deadline algorithm or of a static priority algorithm.

Example 5.7.4: Least slack time scheduling.

To illustrate the use of least slack time scheduling, the algorithm is applied to the task system of Example 5.7.3 for which the attempted earliest deadline schedule appears in Figure 5.29. The resulting processor-sharing schedule is shown in Figure 5.31. The table below each task schedule in the diagram shows the slack time computed for each task at times $t = 0, 1, 2, \cdots$ wherein each time unit may represent a tick of the real time clock. At time $t = 0$, task τ_3 has the least slack time and is provided priority and its own processor.

The other two tasks have the same slack time and share the second processor. This situation does not change until time $t = 32$, at which point, the three tasks share the processors until time $t = 38$. The shaded slack times in the table indicate that the task is already complete, and a second instance has not been released yet. As can be observed, a task can complete with a finite amount of slack time available, which may, indeed, be desirable to prevent missed deadlines due to jitter. An equivalent realizable preemptive schedule is depicted in Figure 5.32. The preemptive schedule is not unique and presents opportunity for rather obvious simplification. For example, by moving some of the task assignments, the schedule of Figure 5.30 can be easily obtained.

The significance of the schedule produced in Figure 5.32, however inefficient in terms of task preemption it might appear, is that the least slack time algorithm produced a valid schedule

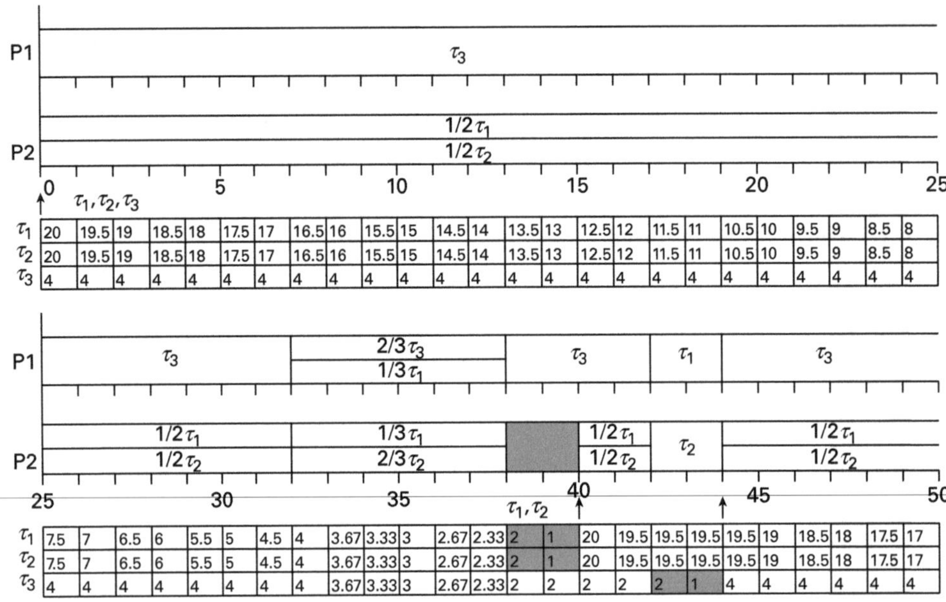

Figure 5.31 Processor Sharing Schedule Produced by Least Slack Time Scheduling Algorithm.

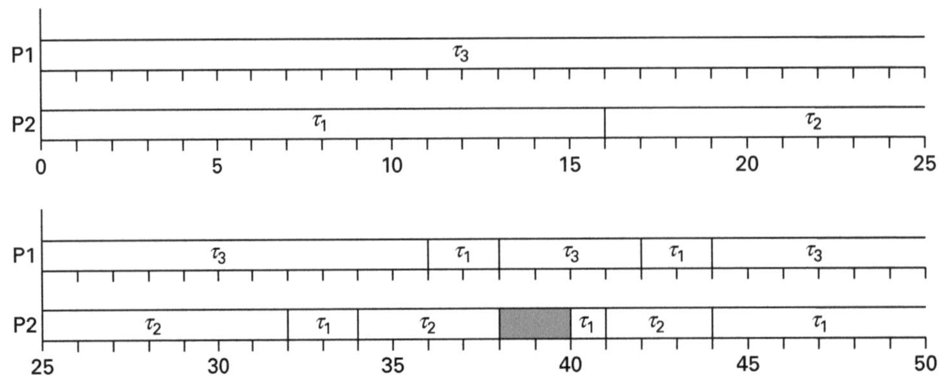

Figure 5.32 Realizable Schedule Produced by Least Slack Time Scheduling Algorithm.

for this task system when the earliest deadline algorithm failed to do so. This result verifies the statement made previously that there are task systems for which the least slack time algorithm can produce valid schedules when the earliest deadline algorithm cannot.

5.8 Summary of Periodic Task Scheduling

The scheduling of periodic tasks is an important component of the design of real time systems software. Most real time systems must interact with the external environment in multiple ways, and the use of concurrently executing tasks is the only means through which concurrent interaction can be obtained in a manner that satisfies the timing constraints of the system. Usually, the real time interaction occurs in a periodic manner, leading to the use of periodic task systems to provide the repetitive processing necessary both to interact with the external environment and to provide the necessary computation. Periodic task systems can be either preemptive or non-preemptive, and this characteristic has a profound effect on the ability to produce task schedules that meet periodic timing requirements.

Periodic task systems usually have some form of synchronization requirements. Synchronization is often in the form of precedence, mutual exclusion, or synchronous communication requirements, usually due to contention for a shared system resource. Even in systems that have a primarily periodic nature, there are nearly always some non-synchronous processing requirements in the form of aperiodic tasks. These aperiodic tasks may have hard or soft deadlines, but, in either case, they affect the periodic schedule, and some accommodation for them must be made. This chapter has presented a discussion of periodic task scheduling, the salient features of which are outlined below.

- In general, a task system can be characterized by specifying for each task: a period, a deadline, and an execution time. If blocking is possible, then a worst-case blocking time must also be specified. Execution times will normally vary from one release of a task to another, a phenomenon known as jitter. As a result, the execution times of tasks are often expressed as worst-case execution times.
- Tasks can be scheduled statically using a cyclic executive approach as discussed in Chapters 4 and 7. Static task schedules are predetermined through the use of task scheduling algorithms that employ the concept of task priority to determine how the various tasks are assigned to the processor(s) as a function of time. The resulting schedule exhibits periodic behavior at a period called the major cycle. Each major cycle is identical to the next. The major cycle is comprised of some number of minor cycles. The processing within different minor cycles is different, and the actual code executed within any given minor cycle is called a frame. Synchronization requirements are met by dividing tasks into subtasks at the synchronization points, and, then, scheduling the subtasks in such a manner that the synchronization requirements are met. Cyclic executive systems are scheduled by periodic, timer-driven interrupts.
- Tasks can be scheduled dynamically through use of a real time multitasking executive (RTOS). Tasks are assigned priorities and are dynamically scheduled as the task system executes by the RTOS, which ensures that the highest priority pending task is always selected for execution. Schedulability analysis is conducted by assuming a fixed execution time for each of the tasks and applying appropriate analysis methods. This is necessary for ensuring that a given task system will meet all deadlines for all task releases. In many situations, the only known analysis method is schedule construction. Synchronization requirements are met through the use of primitive operations, such as semaphores, provided by the real time operating system.
- In general, producing a static schedule for a cyclic executive system or, equivalently, determining if a task system is schedulable using a specified priority assignment in a dynamically scheduled system can be a difficult problem. This fact leads to the use of heuristic methods if the task system does not fall into the rather small category for which efficient techniques for analysis are known.
- Task systems with harmonically related periods greatly simplify the analysis and scheduling processes. In particular, task systems with power of two relationships among the task periods have extremely simple schedulability criteria.

- Arbitrary task preemption increases scheduling flexibility. Arbitrary preemption is a requirement in those few situations in which efficient schedulability analysis is possible.
- Synchronous task systems with arbitrary task preemption exhibit a pseudo-polynomial time algorithm for determining if a particular priority assignment will produce a valid schedule on a single processor. This algorithm consists of constructing the schedule through the longest task period—or of the analytical equivalent of schedule construction.
- In many cases of practical importance, upper bounds on processor utilization relating to schedulability can be formulated and effectively used. Meeting the bound requirement is a sufficient but not necessary condition of schedulability for a task system.
- There are optimal priority assignments for some classes of task systems. An optimal priority assignment is a priority assignment that will produce a valid schedule if one exists.
    - The deadline-monotonic priority assignment is an optimal static priority assignment for synchronous task systems on one processor.
    - The rate-monotonic priority assignment is an optimal static priority assignment for synchronous task systems on one processor when deadlines equal periods.
    - The earliest deadline priority assignment is an optimal dynamic priority assignment for scheduling tasks, synchronous or asynchronous, on one processor.

In many situations of practical importance, the only known algorithm for determining if a given task system is feasible or schedulable is to construct the schedule using a known optimal priority assignment. For scheduling on multiple processors, there is no known optimal priority assignment. Priority assignments used in schedule construction for statically scheduled systems, or those assigned in order to analyze task behavior in a dynamically scheduled system, are used as heuristic approaches to the problem.

CH 6

COMMUNICATION MECHANISMS AND RELIABILITY PROTOCOLS

It is common in a multitasking environment for two or more concurrently executing tasks to share resources, synchronize, and communicate with one another. Shared resources typically include peripherals, memory, data, or a code module such as a subprogram. In many cases, there are also specific synchronization requirements for task sequencing or of task order, task timing, time delays, and passage of program control from one task to another. Most real time operating systems provide mechanisms in the form of data structures, software libraries, and an application programmer interface (API), including reentrant code, semaphores, mailboxes, signals, and timers.

There are potential dangers with all of these features, particularly when implemented in a non-deterministic interwoven real time system. Errors in coding can seriously compromise reliability by the creation of missed deadlines and system deadlock. This chapter discusses the various communication mechanisms typically available in real time systems software and some reliability protocols that have been established over the years to prevent system failure due to errors in communication.

6.1 Reentrant Code

The illustration in Figure 6.1 depicts two concurrently executing tasks, τ_1 and τ_2, both of which make calls to the subroutine in C/C++ (or method in Java and function in Fortran) as they execute. In the scenario depicted, task τ_1 makes a call to the subroutine and so does task τ_2. The arrows indicate the transfer of control upon the subroutine call and upon the return from the subroutine. When a subroutine call is made, parameters are passed. As the subroutine executes, values are assigned to local variables within the subroutine. The block labeled "subroutine local data" represents the storage of these parameters and local variables.

When the return from the subroutine is made, these data are no longer needed and, thus, can be overwritten at the time of the next subroutine call. The problem occurs when a task is suspended

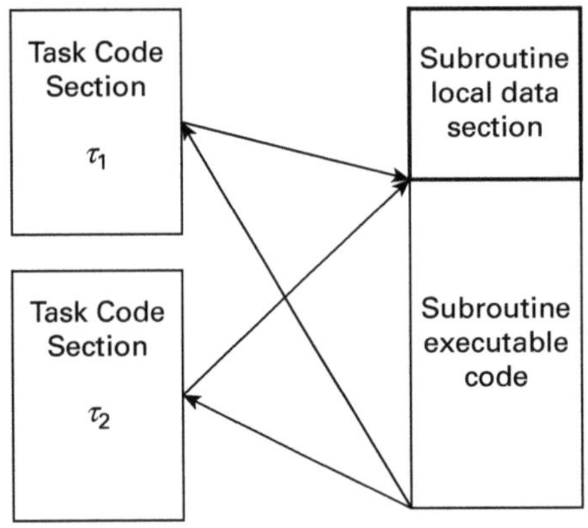

Figure 6.1 Problem with Two Tasks Sharing a Common Resource Like a Subprogram.

while executing in the subroutine, and a second task makes a subroutine call prior to the time the first task resumes execution. The second task will overwrite the local variables generated by the first task and execute the subroutine correctly, but when the first task resumes execution, it will confidently use the current values of the local variables placed there by the second task.

There are two solutions to this problem, one involving *serially reusable code* and the other involving *reentrant code*.

- **Serially reusable code** is code that can be executed by more than one task so long as each task *executes the code to completion* before another task is allowed to enter the code. While a task is executing in serially reusable code module, it is necessary to lock out other tasks. The only restriction applying to serially reusable code is that it must not depend on any specific initial values of the local data.
- **Reentrant code**, on the other hand, is constructed in a manner that allows concurrent execution by more than one task. The basic requirement of reentrant code is that local data needed for execution of the reentrant code is associated with the *user task* rather than with the reentrant code section itself.

Essentially, each user task has its own copy of the local variables, and therefore the local variables are not destroyed when the task is suspended while executing the reentrant code, and another task, subsequently, executes the code. This concept is illustrated in Figure 6.2, which shows the subprogram of the previous example implemented as reentrant code.

6.2 Critical Sections

The synchronization of concurrently executing tasks is an important function of a multitasking executive (RTOS). The common method of synchronizing tasks is the use of the *semaphore*, which is a primitive operation provided by the RTOS itself. In software that uses priority-driven task

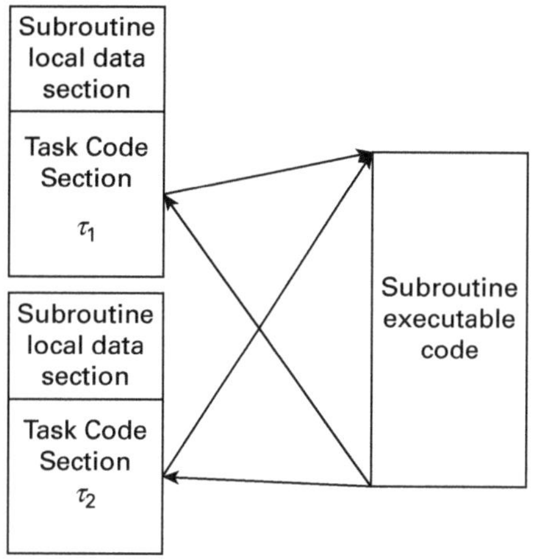

Figure 6.2 Reentrant Code with Local Subprogram Data Storage.

scheduling, task synchronization is required to provide *mutual exclusion* for serially reusable code or to protect a system computational resource such as input/output devices, common data storage, memory, or even a processing unit. The need for mutual exclusion is expressed through the concept of a *critical section*, which is a section of code in a task in which execution cannot occur concurrently with that of a corresponding critical section in any other task. The term *corresponding,* in this context, refers to a common mutual exclusion requirement. This means that corresponding critical sections in a group of tasks define code segments in each task such that while any one of the tasks is executing in its critical section none of the other tasks can be allowed into their corresponding critical sections. In the most common case, a critical section or section of code protects a serially reusable resource.

Figure 6.3 illustrates corresponding critical sections in two tasks, τ_1 and τ_2. Once τ_1 has entered its critical section, τ_2 cannot be allowed to enter its corresponding critical section until τ_1 has completed execution of its critical section. Similarly, if τ_2 has entered its critical section, then, τ_1 cannot be allowed to enter its corresponding critical section until τ_2 has completed execution of its critical section. While the example of Figure 6.3 involves only two tasks, it is quite common for corresponding critical

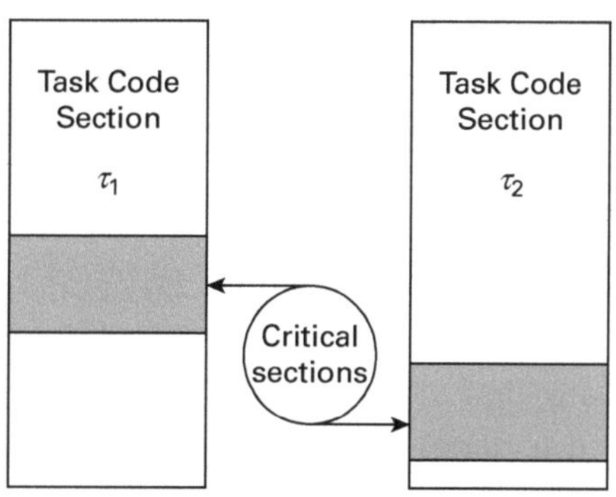

Figure 6.3 Corresponding Critical Sections in Two Tasks.

sections to exist in more than two tasks. Typical programs using a RTOS for task scheduling will have several or many sets of corresponding critical sections, each incorporating a semaphore to provide the required mutual exclusion.

6.3 Semaphores

The *semaphore* is a primitive operator provided by real time operating systems for the purpose of implementing mutual exclusion for serially reusable code or to protect a system resource. The semaphore is represented by an integer variable that can assume only non-negative values. There are only three semaphore operations:

- *INITIALIZE*(S, I)
- *WAIT*(S)
- *SIGNAL*(S)

The *INITIALIZE*(S, I) operation simply sets the value of the semaphore S to the integer value I. The *WAIT*(S) and *SIGNAL*(S) operations are called by concurrently executing tasks. Normally, a task will call *WAIT*(S) upon entering a critical section protected by semaphore S, and it will call *SIGNAL*(S) upon leaving the critical section. The formal definition of the *WAIT*(S) operation is given below.

If $S > 0$*, then* $S = S - 1$*, and the calling task continues execution.*

If $S = 0$*, then* S *is unchanged, and the calling task is suspended.*

The *WAIT*(S) operation is normally called by a task encountering a critical section protected by semaphore S.

The formal definition of the *SIGNAL*(S) operation is given below.

If there is a task that has been suspended by a WAIT(S) *call, then the suspended task is resumed. The calling task continues execution, and the semaphore value remains unchanged.*

If there is no such suspended task, then $S = S + 1$*, and the calling task continues execution.*

If there are multiple tasks that have been suspended by *WAIT*(S) calls, then these tasks form a queue, and one of the queued tasks will be selected for resumption. The queue discipline used to select the task to be resumed is usually first in-first out (FIFO), but more sophisticated multitasking executives may provide other queue disciplines—for example, a priority queue based on task priority. The *SIGNAL*(S) operation is normally called by a task when leaving a critical section protected by semaphore S.

Example 6.3.1: A semaphore protecting corresponding critical sections in three tasks.

An example of a semaphore used to protect corresponding critical sections in three tasks is illustrated in Figure 6.4. Each task, upon encountering its critical section, executes the *WAIT*(*SEM*_1) operation where *SEM*_1 is the name of the semaphore. Upon leaving its critical section, the task executes the *SIGNAL*(*SEM*_1) operation. The three tasks execute asynchronously—a task's execution is not affected by that of the other tasks except when the task is suspended awaiting entry into its critical section. Furthermore, the task scheduling is non-deterministic; it is not under the control of the software developer, but, rather, it is determined dynamically by the scheduling function of the multitasking executive on the basis of assigned priorities. Thus, in order to demonstrate the mutual exclusion produced by use of the semaphore, a scenario must be hypothesized to describe the order in which the tasks arrive at their critical sections. This scenario is presented in the table below.

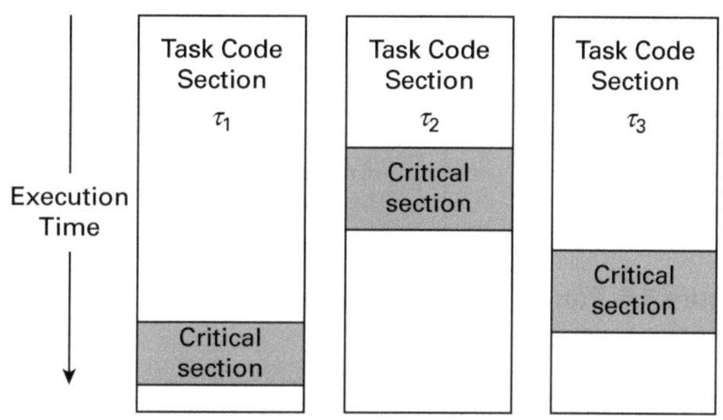

Figure 6.4 Three Tasks with Associated Critical Sections.

After this sequence of actions, all three tasks have transited their associated critical section, and mutual exclusion has been effected. The semaphore *SEM*_1 has been restored to its initializing value of 1. While this scenario presumes a certain sequence of task arrivals at critical sections—task τ_1, task τ_2, and task τ_3, in that order—the reader can verify that any other order of task arrivals will also result in mutual exclusion.

Event	**Call**	**Result**	**Action**
initialization	*INITIALIZE*(*SEM*_1,1)	*SEM*_1 = 1	
τ_1 arrive	*WAIT*(*SEM*_1)	*SEM*_1 remains 0	τ_1 continue
τ_2 arrive	*WAIT*(*SEM*_1)	*SEM*_1 remains 0	τ_2 suspended
τ_3 arrive	*WAIT*(*SEM*_1)	*SEM*_1 remains 0	τ_3 suspended
τ_1 leave	*SIGNAL*(*SEM*_1)	*SEM*_1 remains 0	τ_2 resumed
τ_2 leave	*SIGNAL*(*SEM*_1)	*SEM*_1 remains 0	τ_3 resumed
τ_3 leave	*SIGNAL*(*SEM*_1)	*SEM*_1 = 1	

If a semaphore is initialized to 1, it is called a binary semaphore because the only values it can ever assume are 0 and 1. A binary semaphore allows only one task to be executing in its critical section at any time. The more general form of semaphore is the *counting semaphore*, which, when initialized to have the integer value N, will allow N tasks to execute concurrently in their critical sections. In other words, tasks will be allowed to enter their critical sections until the total is N, after which tasks encountering their critical section will be suspended. Counting semaphores are useful in protecting multiple but limited resources.

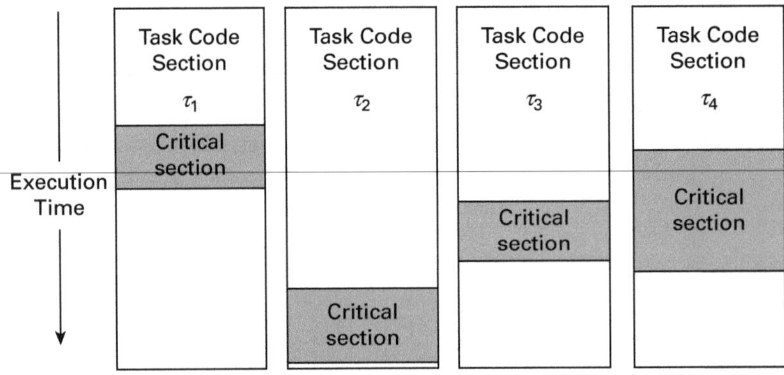

Figure 6.5 Four Tasks with Associated Critical Sections.

Example 6.3.2: A counting semaphore.

Figure 6.5 depicts four tasks with associated critical sections. The tasks use a counting semaphore to permit two tasks to execute in their critical sections concurrently. That is, the semaphore must function to prevent any of the four tasks from entering its critical section while any two of the remaining tasks is executing within its own critical section. The desired result can be obtained by initializing SEM_2 with the value 2.

Again, in order to illustrate the functioning of the semaphore, a scenario of events must be postulated. This scenario is presented in the table below.

Event	**Call**	**Result**	**Action**
initialization	$INITIALIZE(SEM\_2, 2)$	$SEM\_2 = 2$	
τ_1 arrive	$WAIT(SEM\_2)$	$SEM\_2 = 1$	τ_1 continue
τ_2 arrive	$WAIT(SEM\_2)$	$SEM\_2 = 0$	τ_2 continue
τ_3 arrive	$WAIT(SEM\_2)$	SEM_1 remains 0	τ_3 suspended
τ_2 leave	$SIGNAL(SEM\_2)$	SEM_1 remains 0	τ_3 resumed
τ_4 arrive	$WAIT(SEM\_2)$	SEM_1 remains 0	τ_4 suspended
τ_1 leave	$SIGNAL(SEM\_2)$	SEM_1 remains 0	τ_4 resumed
τ_4 leave	$SIGNAL(SEM\_2)$	$SEM\_2 = 1$	
τ_3 leave	$SIGNAL(SEM\_2)$	$SEM\_2 = 2$	

The desired synchronization has been obtained, and *SEM*_2 has been returned to its initializing value. The reader may wish to verify that any other order of task arrivals will result in the satisfaction of the synchronization requirements, and a limit of two tasks will be allowed to concurrently execute in their critical sections.

While the real time operating system performs many functions necessary to support concurrently executing tasks, certainly, the implementation of semaphores is one of the most important. One of the most critical aspects of semaphore implementation is that the *WAIT*(*S*) and *SIGNAL*(*S*) operations should be *atomic* in nature. What this means is that they should never be interruptible by a hardware signal interrupt or any other preemption. It is obvious that such preemption will compromise the validity of the semaphore value and should not be allowed by the definition itself. Practically speaking, most microprocessors used for real time systems have atomic instructions but no provision for atomicity of sets of instructions. Several new processors such as the ARM set of microprocessor cores aimed at the real time market now provide special instructions to read and set a flag or variable. The use of such instructions is tricky and has to be carefully implemented.

Figure 6.6 shows the typical implementation of semaphores to guard serially reusable system resources. A typical semaphore structure implemented in a real time operating system along with function prototypes is shown. Recall that a semaphore may have any number of tasks waiting in its queue of suspended tasks. This list of suspended tasks is maintained by the multitasking executive as a linked list of threads or TCBs. A semaphore is implemented in some RTOSs as a data structure containing only four fields—the name of the semaphore, the current value of the semaphore, and two pointers. One of the pointers points to the first TCB on the linked list of TCBs and the second to the last TCB on the list.

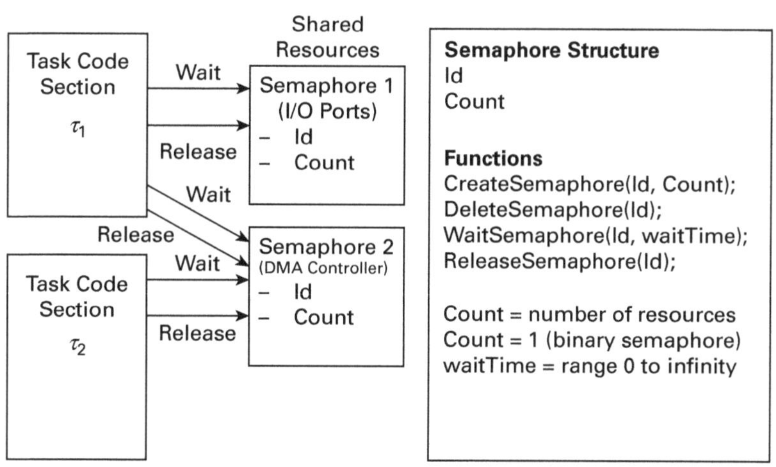

Figure 6.6 Implementation of a Semaphore.

The TCBs are arranged on the list in the order of their resumption—first in-first out, task priority, etc. The pointer to the last TCB is provided so the first in-first out queue discipline can be easily implemented. The reader should bear in mind that the TCBs on the semaphore's list may also be on one or more other linked lists maintained by the multitasking executive. The RTOS may also utilize a state machine to move tasks or threads from different states such as Ready, Suspended, and Running. Some RTOSs provide an additional wait time field that defaults to infinity (wait forever) but can be set to some finite value. This allows the task or thread to exit from the semaphore after waiting for the preset amount of time. This can be quite handy in certain cases to prevent deadlock situations wherein a task or thread will wait forever for a resource. As with any complex program, the real time systems engineer has to carefully study the API for the particular RTOS being used to prevent faulty operation.

6.4 Mailboxes

Another kind of data structure used by multitasking executives for message passing is the *mailbox*. A mailbox differs from an event queue in that it is a system resource that is available to all tasks for the purpose of sending and receiving messages. The implementation of a mailbox might be as illustrated in Figure 6.7. The queue structure used in a modern RTOS is shown with the function prototypes available to the real time developer as a part of the API. Some of the functions may allow the dynamic allocation of memory for the message being passed to other tasks or threads.

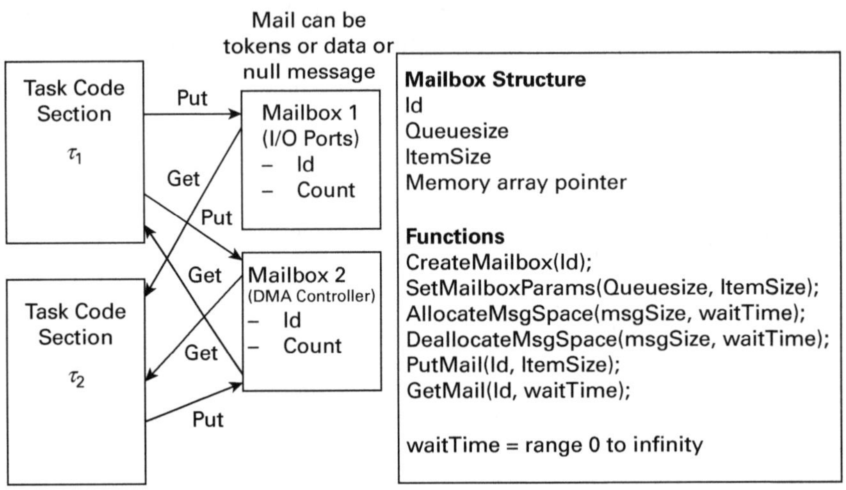

Figure 6.7 Implementation of a Mailbox.

Once a mailbox structure is created, tasks or ISRs can send or put mail in the mailbox. The mail may consist of data or a pointer to a data structure or even something as simple as a token key or flag with no data in it. Another thread or ISR can then read or get the mail from this mailbox structure or queue. Typically, the real time executive will provide API function calls to create a mailbox and a mailbox queue and to deallocate it, put messages in the mail queue, and get messages from the mail queue. The put and get functions can be used from within the task code and typically have timeout values as one of the parameters passed to the function. Thus, a task can wait for a mail message forever or for a specified amount of time. Once the mail is read from the queue, the message is destroyed.

6.5 Queues

Message handling is also an important function of multitasking executives. A message, or *event*, is passed from one task to another in the manner described previously—that is, the sending task must request that the multitasking executive place the event in the event queue of the receiving task, and the receiving task must request that the event be retrieved from the event queue. Each task that receives events must have an event queue in which the events are queued. The implementation using a structure is illustrated in Figure 6.8. Because event queues are protected, only the multitasking executive can access them, thus, forcing both the sending and receiving task to request message handling services from the multitasking executive. The implementation of an event queue implies that any task may queue an event.

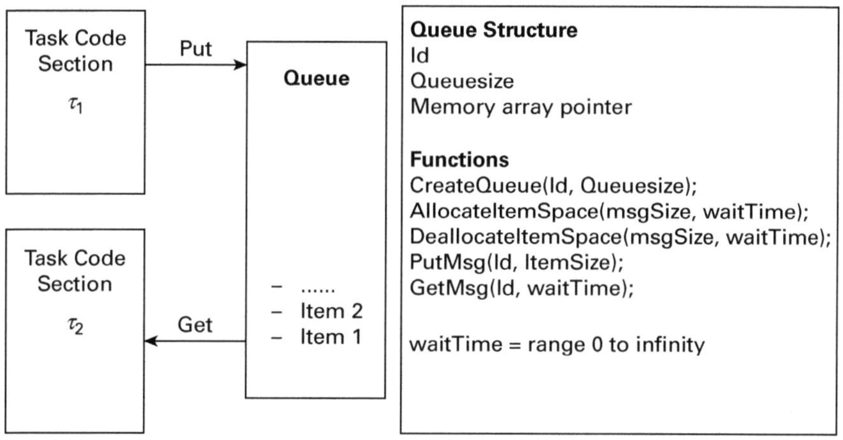

Figure 6.8 Implementation of Message Passing with Queues.

Message queue management functions are typically a part of the real time operating system API. In general, they can control, send, receive, or wait for messages. A message is typically an integer or pointer value that is sent to a task or interrupt service routine. Similar to mailboxes, time-out values are one of the parameters passed to the function. Thus, a task can wait for a message forever or for a specified amount of time. Once the message is read from the queue, the queue pointer is advanced to the next message.

6.6 Other Synchronization Mechanisms—Delays, Timers, and Signals

Most real time executives will provide a number of utilities in order to synchronize multiple tasks, to pass control from one task to another, and to control interaction with peripherals in a timed fashion. These utilities typically can be classified as delays, timers, or signals.

Delays allow a task or peripheral interface code to wait for a specified amount of time or for an unspecified amount of time. The latter is typically a special case of the former with the wait time value set to forever. Delays can be set as a function of the real time clock, and, typically, the value passed to the function call is the number of ticks. These functions can be extremely versatile and allow for fairly precise operation of tasks and peripheral command interfacing.

Timers are typically managed by an API function group that allows creating and controlling of timers and may have callback functions as an additional feature associated with each timer data structure. A *callback function* is called when a time period expires. Both one-shot and periodic timers are possible, depending on the sophistication of the real time executive. A timer can be started, restarted, or stopped and may have a separate task associated with each instance. Callback functions run under control of the timer threads and can be used to trigger events.

Signals are a simple yet effective method of passing control from one task to another or passing specific event information to another task or even the main executive task itself. Typically, the signal API group will allow for a certain number of unique signals that are numbered or enumerated. Each task can, then, *"signal"* (send a signal value) to another task or wait for a specific signal value from another task. This allows for a very convenient passage of control from one task to another.

6.7 Priority Inversion

The schedulability analysis presented in the previous sections has presumed that tasks are independent—that is, there are no synchronization requirements. Real time task systems seldom consist of completely independent tasks. Rather, tasks do interact and must be synchronized. In cyclic executive systems, task synchronization dictates the manner in which tasks and/or subtasks are formed. That is, two tasks that must synchronize at particular points in their execution are divided into subtasks at those synchronization points, and, then, synchronization is effected by corresponding precedence constraints.

In priority-driven, dynamically scheduled systems, synchronization is implemented through the use of semaphores or other primitive operations provided by the real time multitasking executive. It is clear that the use of semaphores and other blocking mechanisms will affect the schedulability of task systems and that the schedulability analysis of systems of independent tasks must be extended if synchronization requirements apply.

The effects of synchronization requirements on scheduling synchronous task systems on one processor and methods for determining task completion times can be studied in terms of worst-case response time, which can be a function of blocking time. The worst-case blocking time b_n represents a conglomerate worst-case blocking of task τ_n by all lower priority tasks. Obviously, one primary source of blocking by lower priority tasks is that caused by task synchronization. While techniques that are effective in assessing the effects of the worst-case blocking time for any given task exist, there is a serious question of how these worst-case blocking times can be evaluated. A simple example serves to illustrate the potential problems that may be encountered.

Depicted in Figure 6.9 is a scenario in which three tasks τ_1, τ_2, and τ_3, with descending order of priority—are concurrently executing. Tasks τ_1 and τ_3 share a common resource protected by semaphore S. Consequently, tasks τ_1 and τ_3 have corresponding critical sections of code, and each task executes

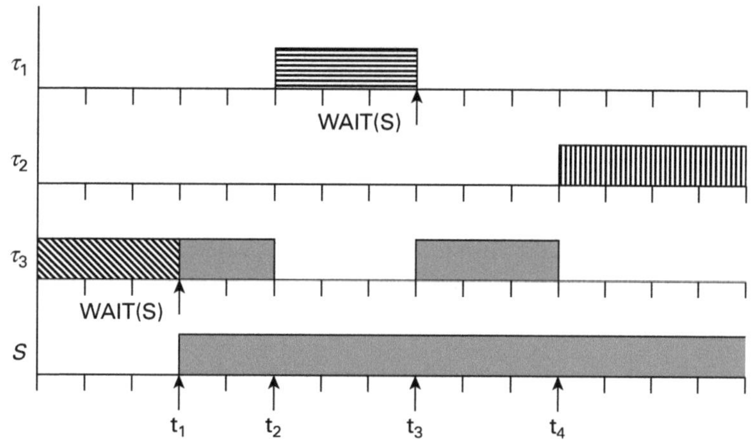

Figure 6.9 An Example of Priority Inversion.

WAIT(*S*) prior to entering the critical section and executes *SIGNAL*(*S*) upon leaving. In the scenario depicted in the figure, the following events occur in the order shown:

- At time $t = t_1$, task τ_3 executes *WAIT*(*S*) and enters its critical section. Semaphore *S* is locked as indicated in the diagram by the fill pattern of the semaphore. Also indicated in the diagram is a change in the fill pattern of task τ_3 to match that of semaphore S in order to indicate that task τ_3 holds the lock on semaphore *S*.
- At time t_2, task τ_3 is preempted by the high priority task τ_1, which then begins to execute.
- At time t_3, task τ_1 encounters its critical section and executes *WAIT*(*S*). Since semaphore *S* is currently locked by task τ_3, task τ_1 is suspended. Task τ_3 then resumes execution. Task τ_1 is now blocked by the lower priority task τ_3, and the blocking will last until task τ_3 completes its critical section and executes *SIGNAL*(*S*).
- At time $t = t_4$, task τ_3 is preempted by task τ_2, which then begins to execute.
- At this point, task τ_1 is blocked by the lower priority task τ_3—awaiting task τ_3's execution of *SIGNAL*(*S*) upon leaving its critical section, but task τ_3 is not executing.

Rather, it is suspended within its critical section due to circumstances entirely unrelated to the particular resource protected by semaphore *S*. In effect, high priority task τ_1 is being preempted by the lower priority task τ_3. This phenomenon is termed *priority inversion* and is an important concern in the functioning of priority-driven, dynamically scheduled systems. Delays in the completion of high priority tasks caused by priority inversion are *unpredictable* and can be *arbitrarily long*.

From the point of view of schedulability analysis, priority inversion is a serious concern for the following reason. Schedulability analysis for synchronous task systems that are preemptable, that use static priorities, and that are subject to blocking effects include the following approaches.

1. Using a utilization bound to establish a sufficient but not necessary condition.
2. Constructing the schedule through the longest period.
3. Determining the time of first completion for all tasks.
4. A combination of (1) and (3) above.

All these methods require the specification of the worst-case blocking time for each task in the task system. *In the presence of priority inversion, blocking time can be unbounded.* In order to use these analysis techniques—and, more importantly, to be able to use priority-driven, dynamic scheduling in the presence of synchronization effects that may be subject to priority inversion—some technique for limiting the effects of priority inversion must be used. Since priority inversion is a result of the method by which task synchronization is implemented, its effect can be limited only through some basic modification to the manner in which task synchronization is implemented. This requirement leads to the introduction of *synchronization protocols*.

> *A synchronization protocol is a protocol for two tasks including an implementation primitive(s), a policy for deciding what to do if the synchronization cannot take place and a policy for deciding what priority level to assign while the tasks remain synchronized.*

The simplest synchronization protocol is to always grant a requesting task a lock on an unlocked semaphore and to have the task execute the critical section at its normally assigned priority. This was the case in the scenario depicted in Figure 6.9 and, as illustrated, it can lead to unbounded priority inversion.

One other simple approach to preventing priority inversion is to protect critical sections from preemption—termed the *highest locker's protocol* because it is implemented by having tasks execute their critical sections at the priority of the highest priority task that shares the semaphore. While this approach may be appropriate for very short critical sections, it can lead to unnecessary blocking of high priority tasks and, therefore, may not be appropriate for longer critical sections. For example, once a low priority task enters a long, non-preemptable critical section that protects a shared resource, a high priority task that does not require access to the resource may be unnecessarily blocked for the duration of execution of the critical section by the lower priority task. A better approach to the prevention of priority inversion is the use of one of several related techniques known as the *priority inheritance protocols*. The following section describes these synchronization protocols.

6.8 Priority Inheritance Protocol

While there may be several ways in which a lower priority task can block a higher priority task, the most prevalent concept is that of a semaphore. The discussion of the priority inheritance protocols will be presented in terms of the semaphore and the critical sections that it protects. The basic concept underlying the priority inheritance protocols is the following.

> *When a task blocks one or more higher priority tasks, it executes its critical section at the priority of the highest priority task that it blocks. That is, it inherits the priority of the highest priority task that it blocks. When leaving the critical section, the task reverts to its original priority.*

The concept is fairly straightforward—by executing at the higher priority of the blocked task, the lower priority task will not be preempted by any task with priority lower than that of the highest priority blocked task. Priority inversion will, thus, be avoided without unnecessary blocking of yet higher priority tasks.

The discussion to follow requires the definition of some basic terminology.

- A task is said to be *blocked* when it is forced to wait for the execution of a *lower* priority task. When a task is forced to wait for the execution of a *higher* priority task, it is not considered to be blocked.
- The $j-th$ critical section in task τ_i that can block task τ_n is referenced as C_{nij}
- The semaphore protecting critical section C_{nij} is referenced as S_{nij}
- A task τ_n is said to hold the lock on semaphore S if it has successfully executed $WAIT(S)$.
- A task τ_n is said to be blocked by the critical section C_{nij} if it has a higher priority than task τ_i but is required to wait until task τ_i completes critical section C_{nij} before it can resume execution.
- A task τ_n is said to be *blocked by task* τ_i *through semaphore* S if the critical section C_{nij} blocks task τ_n and $S = S_{nij}$.
- When critical sections are nested—that is, when critical sections are contained within other critical sections—the outer critical section is said to enclose the inner critical sections. If there is no nesting, a critical section is considered to be an enclosing critical section.

To illustrate the behavior of task scheduling employing the basic priority inheritance protocol, the scenario depicted in Figure 6.9 will be examined but, now, with the basic priority inheritance protocol in effect. The scenario is one in which three tasks—τ_1, τ_2, and τ_3, with descending order of priority—concurrently execute. Tasks τ_1 and τ_3 share a common resource protected by semaphore S. The result of synchronizing the tasks using semaphore S and the basic priority inheritance protocol is shown in Figure 6.10. The result differs from that of Figure 6.9 in that task τ_2 does not preempt the lower priority task τ_3 at time $t = 4$ because task τ_3 executes its critical section at the inherited priority of task τ_1, which it is blocking. The result is that the high priority task τ_1 is delayed only for the time required for task τ_3 to complete its critical section. There is no priority inversion effect.

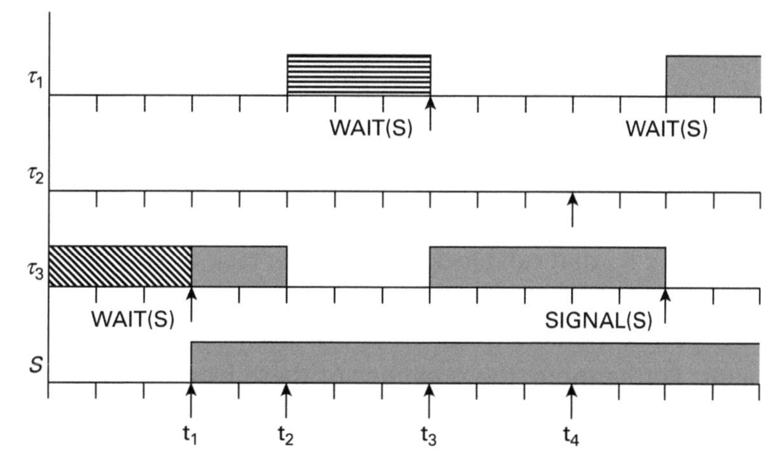

Figure 6.10 Priority Inversion Avoided Through the Use of the Priority Inheritance Protocol.

While this example illustrates the concept of priority inheritance, it represents the most basic kind of scenario in which the priority inheritance protocol might be applied. For a more complete understanding, the protocol must be expanded and formally defined in the following manner.

The highest priority-assigned or inherited task always runs unless it is blocked.

Standard binary semaphore concepts apply.

A task executes at its assigned priority unless it is executing in a critical section and blocks one or more higher priority tasks, in which case, it executes at its inherited priority, which is the priority of the highest priority blocked task.

Priority inheritance is transitive—if task τ_3 blocks task τ_2 and task τ_2 blocks task τ_1, then, task τ_3 inherits the priority of task τ_1.

The basic priority inheritance protocol, as described above, permits a high priority task to be blocked by a low priority task in two situations.

- *Direct blocking* of task τ_n occurs when it attempts to lock a semaphore—executes *WAIT*(*S*)—that is currently locked by some lower priority task. This is the kind of blocking normally expected when semaphores or other blocking mechanisms are used to synchronize concurrently executing tasks.
- *Push-through blocking* occurs when a task is preempted by a lower priority task that is currently executing at a higher inherited priority. Push-through blocking is a direct result of the use of priority inheritance and is the price paid for the avoidance of priority inversion.

Use of the basic priority inheritance protocol results in task scheduling with two properties that serve to bound the time that tasks can be blocked by other tasks of lower priority.

1. A higher priority task τ_H can be blocked by a lower priority task τ_L for, at most, the duration of one enclosing critical section C_{HLi} regardless of the number of semaphores shared by tasks τ_H and τ_L. This property clearly establishes a bound on the time that task τ_L can block task τ_H as the duration of the longest critical section in task τ_L that can block task τ_H.
2. If there are I semaphores S_{nij} that can block task τ_n, then task τ_n can be blocked at most I times.

These properties clearly permit the determination of the worst-case blocking times required as part of schedulability analysis. Furthermore, the durations of these blocking times are bounded. For any particular task τ_n, it is necessary only to identify the critical sections C_{nij} in lower priority tasks τ_i that require the longest execution time. By summing these execution times, the worst-case blocking time for task τ_n is obtained.

6.9 System Risk Conditions

Even though the basic priority inheritance protocol solves the problem of unbounded priority inversion, there remain some undesirable properties that, as it turns out, can be avoided. These properties are listed below.

- The basic priority inheritance protocol does not prevent *deadlock*. Deadlock can occur in dynamically scheduled systems when blocking mechanisms such as semaphores are employed, and it arises from a situation in which each of two tasks is blocked by the other task, with the result that neither task can execute. The actual blocking may occur through interactions with several other tasks, but the basic cause involves two mutually blocking tasks. The properties listed above are valid only if some means is employed to ensure that deadlock cannot occur.
- The blocking time of a task, even though it is bounded, can be large due to sequential blocking by two or more semaphores. This phenomenon—called *chain blocking*—is clearly permitted by property (2) above and, in fact, is characteristic of the basic priority inheritance protocol.
- *Transitive blocking*—the situation in which task τ_i blocks high priority task τ_n and, in turn, is blocked by task τ_j—can occur in the presence of the basic priority inheritance protocol. In effect, task τ_n is then blocked by task τ_j.

6.9.1 Deadlock

The occurrence of deadlock in the presence of the basic priority inheritance protocol is demonstrated by the scenario depicted in Figure 6.11, showing the two tasks τ_2 and τ_3 sharing the two semaphores S_2 and S_3. The sequence of events occurring in this scenario are listed below.

- At time $t = t_1$, low priority task τ_3 executes $WAIT(S_3)$ and enters the critical section protected by semaphore S_3, thus, locking semaphore S_3. The diagram of Figure 6.11 indicates this occurrence by the change in the fill pattern of task τ_3 to match that of semaphore S_3.
- At time $t = t_2$, Task τ_3 is preempted by the higher priority task τ_2, which begins execution.
- At time $t = t_3$, task τ_2 executes $WAIT(S_2)$ and enters its critical section protected by semaphore S_2, thus, locking semaphore S_2. The diagram of Figure 6.11 indicates this occurrence by the change in the fill pattern of task τ_2 to match that of semaphore S_2.
- At time $t = t_4$, task τ_1 preempts task τ_2 and begins execution.
- At time $t = t_5$, task τ_1 executes $WAIT(S_1)$ and enters its critical section protected by semaphore S_1, thus, locking semaphore S_1. The diagram of Figure 6.11 indicates this occurrence by the change in the fill pattern of task τ_1 to match that of semaphore S_1.
- At time $t = t_6$, task τ_1 executes SIGNAL(S_1) and leaves its critical section protected by semaphore S_1, thus, unlocking locking semaphore S_1. The diagram of Figure 6.11 indicates this occurrence by the change in the fill pattern of task τ_1 to its original form.
- At time $t = t_7$, task τ_1 completes, permitting task τ_2 to resume execution.
- At time $t = t_8$, task τ_2 executes $WAIT(S_3)$ and is suspended because semaphore S_3 is locked by task τ_3. Task τ_3, then, resumes execution at the inherited priority of task τ_2.

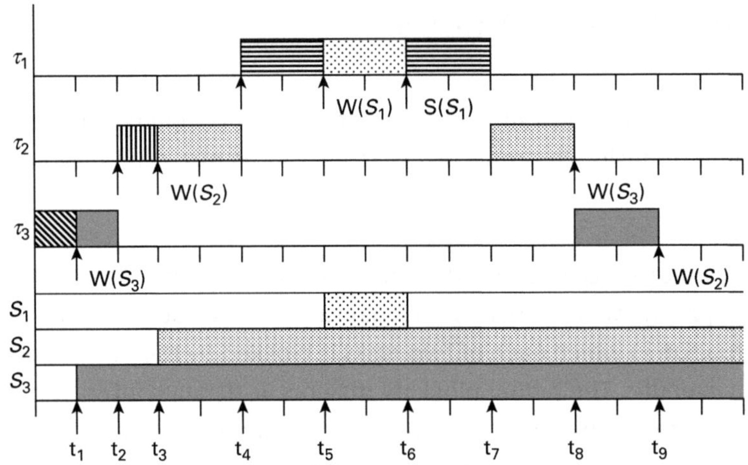

Figure 6.11 Example of Deadlock in a Task Schedule Employing the Basic Priority Inheritance Protocol.

- At time $t = t_9$, task τ_3 executes *WAIT*(S_2) and is suspended because semaphore S_2 is currently locked by task τ_2.

The situation at time $t = t_9$ is one of mutual blocking—task τ_2 is waiting for task τ_3 to complete its critical section and execute *SIGNAL*(S_2) while, at the same time, task τ_3 is waiting for task τ_2 to complete its critical section and execute *SIGNAL*(S_3). Deadlock situations such as this can occur in any dynamically scheduled system using blocking mechanisms such as semaphores whether or not the basic priority inheritance is employed. The purpose of this example is to demonstrate the fact that use of the basic priority inheritance protocol does not prevent deadlock situations from arising. If deadlock is possible, some technique for avoiding it must be used. Such techniques include imposing an ordering of semaphore accesses.

6.9.2 Chain Blocking

Chain blocking is the sequential blocking of a high priority task by more than one semaphore. This situation is characteristic of the basic priority inheritance protocol; although a single lower priority task can block a high priority task τ_n for at most the duration on its longest critical section, *every* lower priority task may block task τ_n. The result can be substantial blocking of a high priority task. A simple example of the occurrence of chain blocking under the basic priority inheritance protocol is illustrated in Figure 6.12. Chain blocking of the high priority task τ_1 and lower priority tasks τ_2 and τ_3 occurs from time $t = t_5$ to time $t = t_8$, as indicated in the diagram of Figure 6.12. The sequence of events occurring in this scenario is described below.

- At time $t = t_1$, task τ_3 executes *WAIT*(S_1) and enters its critical section, thus, locking semaphore S_1.
- At time $t = t_2$, task τ_2 preempts task τ_3.
- At time $t = t_3$, task τ_2 executes *WAIT*(S_2) and enters its critical section, thus, locking semaphore S_2.

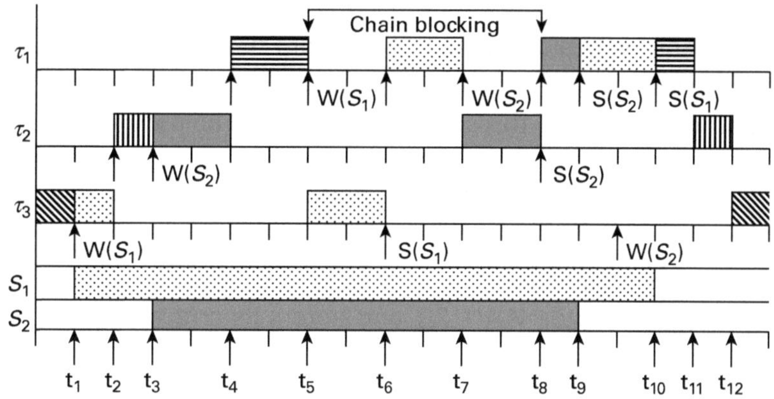

Figure 6.12 Example of Chain Blocking in a Schedule Employing the Priority Inheritance Protocol.

- At time $t = t_4$, task τ_1 preempts task τ_2.
- At time $t = t_5$, task τ_1 executes *WAIT*(S_1). Since semaphore S_1 is currently locked by task τ_3, task τ_1 is suspended. In accordance with the basic priority inheritance protocol, task τ_3 inherits the priority of task τ_1 and resumes execution.
- At time $t = t_6$, task τ_3 executes *SIGNAL*(S_1) and releases the lock on semaphore S_1. Task τ_3 reverts to its original priority and task τ_1 resumes execution, immediately locking semaphore S_1 and executing into its critical section.
- At time $t = t_7$, task τ_1 executes *WAIT*(S_2). Since semaphore S_2 is currently locked by task τ_2, task τ_1 is suspended. In accordance with the basic priority inheritance protocol, task τ_2 inherits the priority of task τ_1 and resumes execution.
- At time $t = t_8$, task τ_2 executes *SIGNAL*(S_2) and releases the lock on semaphore S_1. Task τ_2 reverts to its original priority and is preempted by task τ_1, which immediately executes *WAIT*(S_2), locks semaphore S_2, and executes into its inner critical section.
- At time $t = t_9$, task τ_1 executes *SIGNAL*(S_2) and releases the lock on semaphore S_1. Task τ_1 continues execution in its outer critical section.
- At time $t = t_{10}$, task τ_1 executes *SIGNAL*(S_1) and releases the lock on semaphore S_1. Task τ_1 continues execution.
- At time $t = t_{12}$, task τ_2 completes. Task τ_3 resumes execution.

The chain blocking can be seen occurring from time $t = t_5$ to time $t = t_8$, when task τ_1 is blocked first by task τ_3 (from $t = t_5$ to time $t = t_6$), and, then, by task τ_2 (from time $t = t_7$ to time $t = t_8$). In a system with many tasks, the highest priority tasks may be subject to chain blocking by any number of lower priority tasks, creating the possibility of substantial blocking delay.

6.9.3 Transitive Blocking

Transitive blocking occurs when a high priority task is blocked by a lower priority task that is, in turn, blocked by an even lower priority task. A simple example of the occurrence of transitive blocking under the basic priority inheritance protocol is illustrated by the scenario depicted in Figure 6.13 where the high priority task τ_1 is transitively blocked by task τ_3 from time $t = t_6$ to time $t = t_7$. The blocking occurs while task τ_1 is blocked by task τ_2, which is, in turn, blocked by task τ_3.

The sequence of events occurring in this scenario is described below.

- At time $t = t_1$, task τ_3 executes *WAIT*(S_1) and enters its critical section, thus, locking semaphore S_1.
- At time $t = t_2$, task τ_2 preempts task τ_3.
- At time $t = t_3$, task τ_2 executes *WAIT*(S_2) and enters its critical section, thus, locking semaphore S_2.
- At time $t = t_4$, task τ_2 executes *WAIT*(S_1). Since semaphore S_1 is currently locked by task τ_3, task τ_2 is suspended. In accordance with the provisions of the basic priority inheritance protocol, task τ_3 inherits the priority of task τ_2 and resumes execution.
- At time $t = t_5$, task τ_1 preempts task τ_3.
- At time $t = t_6$, task τ_1 executes *WAIT*(S_2). Since semaphore S_2 is currently locked by task τ_2, task τ_1 is suspended. In accordance with the provisions of the basic priority inheritance protocol—which, in this situation, requires transitive inheritance—task τ_3 inherits the priority of task τ_1 and resumes execution. Task τ_1 is now transitively blocked.
- At time $t = t_7$, task τ_3 executes *SIGNAL*(S_1) and reverts to its original priority. In accordance with the provisions of the basic priority inheritance protocol task, τ_2 now inherits the priority of task τ_1 and resumes execution, immediately locking semaphore S_1. At this point, task τ_2 is executing in nested critical sections and holds the locks on both semaphore S_1 and S_2. The transitive blocking of task τ_1 by task τ_3 has now ended.
- At time $t = t_8$, task τ_2 executes *SIGNAL*(S_1), thus, unlocking semaphore S_1, and continues execution at the priority of task τ_1.
- At time $t = t_9$, task τ_2 executes *SIGNAL*(S_2), thus, unlocking semaphore S_2, and reverts to its original priority. Task τ_1 preempts task τ_2, immediately locks semaphore S_2, and continues execution in its critical section.
- At time $t = t_{10}$, task τ_2 executes *SIGNAL*(S_2), thus, unlocking semaphore S_2, and continues execution.

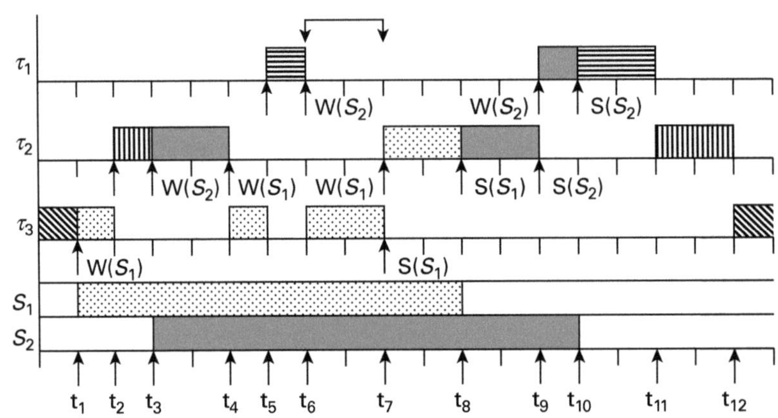

Figure 6.13 Example of Transitive Blocking in a Schedule Employing the Basic Priority Inheritance Protocol.

- At time $t = t_{11}$, task τ_1 completes, permitting task τ_2 to resume execution.
- At time $t = t_{12}$, task τ_2 completes, permitting task τ_3 to resume execution.

The transitive blocking of task τ_1 from time $t = t_6$ to time $t = t_7$ is illustrative of the manner in which this kind of indirect blocking can occur whether or not the basic priority inheritance protocol is employed. This example is used to illustrate the fact that the basic priority inheritance protocol does not prevent transitive blocking.

6.10 Priority Ceiling Protocol

The examples of the previous section have shown that the basic priority inheritance protocol—while effective in limiting the duration of priority inversion effects—does not prevent deadlock, does permit chain blocking, and does not prevent transitive blocking. An improved synchronization protocol that addresses these three issues is the *priority ceiling protocol*. The underlying concept of the priority ceiling protocol is stated below.

> *When a task τ_n preempts the critical section of another task and then attempts to execute one of its own critical sections, it will be suspended unless its critical section can be guaranteed to execute at a priority higher than the inherited priorities of all currently preempted critical sections.*

The goal here is to prevent the formation of deadlock conditions and to prevent both chain blocking and transitive blocking. The implementation of the priority ceiling protocol uses the concept of a *priority ceiling* for each semaphore, as outlined below.

- Basic priority inheritance is in effect—a task inherits the priority of the highest priority task it is blocking.
- Each semaphore is assigned a *priority ceiling*, which is the value of the priority of the highest priority task that shares the semaphore.
- A task is permitted to execute *WAIT*(*S*) and enter a critical section only if its *current* priority is *higher than* the priority ceilings of all semaphores currently locked by *other tasks*.

Since basic priority inheritance is in effect, a task may be executing either with its assigned priority or with an inherited priority. The current priority is the priority at which the task is currently executing. This protocol leads to the use of the term *ceiling blocking* to describe the suspension of a task—attempting to execute *WAIT*(*S*) and enter a critical region—because its current priority does not exceed the priority ceilings of all semaphores currently locked by other tasks.

The priority ceiling protocol exhibits the following properties, which represent considerable improvement over those of the basic priority inheritance protocol.

- The priority ceiling protocol prevents deadlock. Conditions leading to deadlock through mutual blocking do not arise.
- The priority ceiling protocol prevents chain blocking. While the basic priority inheritance protocol allows a task to be blocked by *each* lower priority task for, at most, the duration of one enclosing

critical section, thus, producing the chain blocking effect, the priority ceiling protocol permits a task to be blocked for, at most, the duration of a single enclosing critical section. That is, a task can be blocked by only a single lower priority task.

- The priority ceiling protocol prevents transitive blocking.

The formal definition of the priority ceiling protocol is given below.

The highest priority—assigned or inherited—task always executes unless it is blocked.

A task executes at its assigned priority unless it is executing in a critical section and blocks one or more higher priority tasks, in which case, it executes at its inherited priority, which is the priority of the highest priority blocked task (basic priority inheritance).

Priority inheritance is transitive.

Standard binary semaphore concepts apply, except in the case that a task attempting to lock a semaphore and enter the corresponding critical section will be suspended and the semaphore lock denied if the task's current priority is not higher than the priority ceiling of all semaphores currently locked by other tasks (ceiling blocking).

6.10.1 Prevention of Deadlock

To illustrate the effectiveness of the priority ceiling protocol in preventing deadlock, the scenario depicted in Figure 6.11 is repeated in Figure 6.14, except that now the priority ceiling protocol is in effect. The priority ceiling of a semaphore is the priority of the highest priority task sharing the semaphore. In this case, semaphore S_1 is shared by task τ_1 and other tasks not shown in the diagram. Semaphores S_2 and S_3 are shared by tasks τ_2 and τ_3. Thus, semaphore S_1 has a priority ceiling equal to the priority

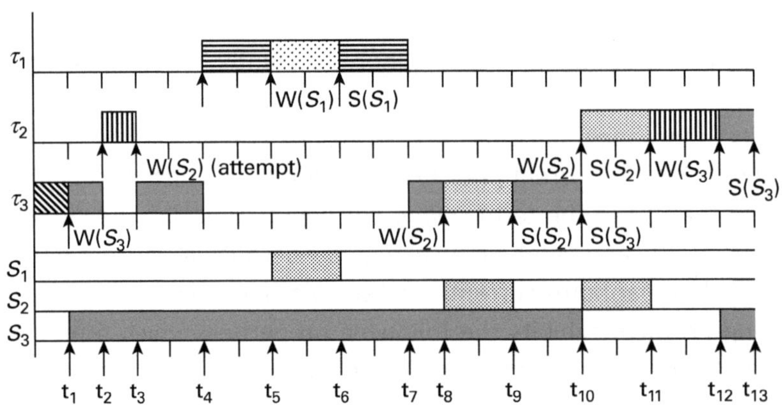

Figure 6.14 Prevention of Deadlock by the Priority Ceiling Protocol (S_1 Priority Ceiling = Priority of Task τ_1 and S_2, S_3 Priority Ceiling = Priority of Task τ_2).

of task τ_1 while tasks τ_2 and τ_3 have a priority ceiling equal to the priority of task τ_2. The deadlock situation is avoided at time $t = t_3$ when task τ_2 is ceiling blocked.

The sequence of events occurring in this scenario is described below.

- At time $t = t_1$, task τ_3 executes *WAIT*(S_3), thus, locking semaphore S_3, and executes into its critical section.
- At time $t = t_2$, task τ_2 preempts task τ_3 and begins execution.
- At time $t = t_3$, task τ_2 attempts to lock semaphore S_2 by executing *WAIT*(S_2) but is prevented from doing so by the provisions of the priority ceiling protocol—task τ_2 is currently executing at its assigned priority, which *does not exceed* the priority ceiling of the currently locked semaphore S_3. Task τ_3 then resumes execution at the inherited priority of task τ_2. Task τ_2 is *ceiling blocked* at this point and not allowed to lock semaphore S_2, and it is this blocking that eventually avoids the deadlock condition.
- At time $t = t_4$, task τ_1 preempts task τ_3 and begins execution.
- At time $t = t_5$, task τ_1 executes *WAIT*(S_1), locks semaphore S_1, and enters its critical section. In accordance with the provisions of the priority ceiling protocol, task τ_1 is permitted the lock on semaphore S_1 because its priority is higher than the priority ceiling of the currently locked semaphore S_3.
- At time $t = t_6$, task τ_1 executes *SIGNAL*(S_1) and leaves its critical section.
- At time $t = t_7$, task τ_1 completes. Task τ_3 resumes execution, still, at the inherited priority of task τ_2.
- At time $t = t_8$, task τ_3 executes *WAIT*(S_2), thus, locking semaphore S_2, and executes into its inner critical section. Task τ_3 is permitted to lock semaphore S_2 because there are no currently locked semaphores.
- At time $t = t_9$, task τ_3 executes *SIGNAL*(S_2), thus, unlocking semaphore S_2, and continues execution in its outer critical section. Task τ_2 remains ceiling blocked because of task τ_3's lock on semaphore S_3.
- At time $t = t_{10}$, task τ_3 executes *SIGNAL*(S_3), thus, unlocking semaphore S_3, and now reverts to its assigned priority. Task τ_2 is no longer ceiling blocked and immediately preempts task τ_3, executes *WAIT*(S_2) —thus, locking semaphore S_2—and executes into its critical section.
- At time $t = t_{11}$, task τ_2 executes *SIGNAL*(S_2), thus, unlocking semaphore S_2, and continues execution.
- At time $t = t_{12}$, task τ_2 executes *WAIT*(S_3), thus, locking semaphore S_3, and executes into its second critical section.
- At time $t = t_{13}$, task τ_2 executes *SIGNAL*(S_3), thus, unlocking semaphore S_3, and continues execution.

Comparing the diagram of Figure 6.14 with that of Figure 6.11 indicates that the potential deadlock is avoided at time $t = t_3$ when task τ_2 is denied the lock on semaphore S_2. It is also noted that task τ_1 is never ceiling blocked because its priority is higher than the priority ceilings of semaphores S_2 and S_3. Furthermore, task τ_2 is blocked from $t = t_3$ to t_4 and from $t = t_7$ to t_{10}, which is the duration of one enclosing critical section in the lower priority task τ_3. Indeed, one of the properties of the priority ceiling protocol is that a task can be blocked by at most the duration of a single enclosing critical section.

This property is further illustrated in the following example scenario, which illustrates the prevention of chain blocking through use of the priority ceiling protocol.

6.10.2 Prevention of Chain Blocking

Figure 6.12 depicts a scenario in which high priority task τ_1 is chain blocked by lower priority tasks τ_2 and τ_3 when the basic priority inheritance protocol is in effect. The same scenario, but now with the priority ceiling protocol in effect, is shown in the diagram of Figure 6.15. Again, it is the denial of the lock on semaphore S_2 at time $t = t_3$ that prevents the blocking of the high priority task τ_1 for no more than the duration of one enclosing critical section—this one in lower priority task τ_3.

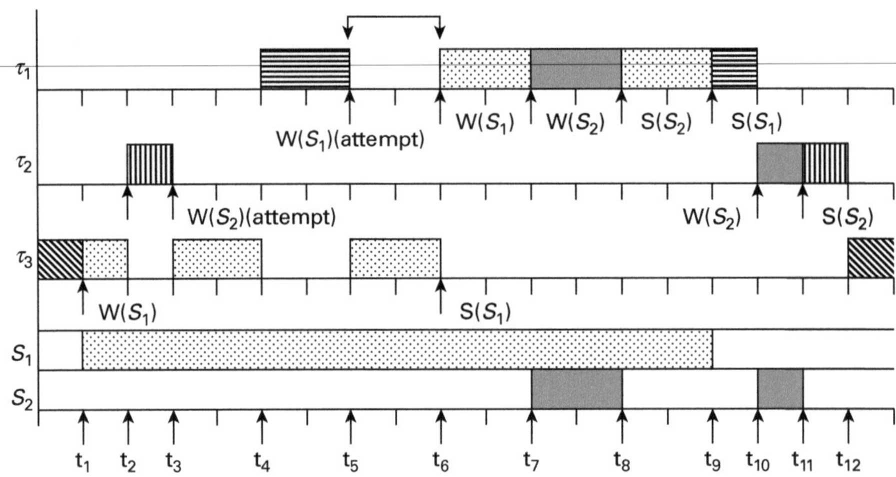

Figure 6.15 Prevention of Chain Blocking by the Priority Ceiling Protocol (S_1, S_2 Priority Ceiling = Priority of Task τ_1).

The sequence of events occurring in this scenario is described below.

- At time $t = t_1$, task τ_3 executes *WAIT*(S_1), thus, locking semaphore S_1, and executes into its critical section.
- At time $t = t_2$, task τ_2 preempts task τ_3 and begins execution.
- At time $t = t_3$, task τ_2 attempts to lock semaphore S_2 by executing *WAIT*(S_2) but is prevented from doing so by the provisions of the priority ceiling protocol—task τ_2 is currently executing at its assigned priority, which *does not exceed* the priority ceiling of the currently locked semaphore S_1. Task τ_3 then resumes execution at the inherited priority of task τ_2. Task τ_2 is ceiling blocked at this point and not allowed to lock semaphore S_2.
- At time $t = t_4$, task τ_1 preempts task τ_3 and begins execution.
- At time $t = t_5$, task τ_1 attempts to execute *WAIT*(S_1). In accordance with the provisions of the priority ceiling protocol, task τ_1 is denied the lock on semaphore S_1 because its priority is not higher than the priority ceiling of the currently locked semaphore S_1. Task τ_1 is ceiling blocked. Task τ_3 resumes execution at the inherited priority of task τ_1.
- At time $t = t_6$, task τ_3 executes *SIGNAL*(S_1), thus, unlocking semaphore S_1, and reverts to its assigned priority. Since no semaphores are currently locked, tasks τ_1 and τ_2 are no longer

ceiling blocked. Task τ_1 then preempts task τ_3, executes *WAIT*(S_1) and locks semaphore S_1, and executes into its critical section.

- At time $t = t_7$, task τ_1 executes *WAIT*(S_2). Since task τ_1 itself holds the lock on semaphore S_1, it is permitted to lock the semaphore and execute into its inner critical section.
- At time $t = t_8$, task τ_1 executes *SIGNAL*(S_2), thus, unlocking semaphore S_2, and exits its inner critical section.
- At time $t = t_9$, task τ_1 executes *SIGNAL*(S_1), thus, unlocking semaphore S_1, and exits its outer critical section.
- At time $t = t_{10}$, task τ_1 completes, allowing task τ_2 to resume execution. Task τ_2 immediately locks semaphore S_2 and enters its critical section.
- At time $t = t_{11}$, task τ_2 executes *SIGNAL*(S_2), thus, unlocking semaphore S_2, and exits its critical section.
- At time $t = t_{12}$, task τ_2 completes, permitting task τ_3 to resume execution.

The high priority task τ_1 is blocked only for a portion of the duration of the critical section of task τ_3—from time $t = t_5$ to t_6. Task τ_2 is blocked, as well, but only for a portion of the duration of the critical section of task τ_3, from time $t = t_3$ to t_4 and from time $t = t_5$ to t_6. For the remainder of the duration from time $t = t_3$ to t_{10}, it is preempted by task τ_1.

The priority ceiling protocol is, thus, seen to prevent the chain blocking effect present in the scenario of Figure 6.12, which is produced even with the basic priority inheritance protocol in effect.

6.10.3 Prevention of Transitive Blocking

The third improvement over the basic priority inheritance protocol provided by the priority ceiling protocol is the prevention of transitive blocking such as that illustrated by the scenario depicted in Figure 6.13. This same scenario, scheduled now with the priority ceiling protocol in effect, is shown in Figure 6.16. The sequence of events occurring in this scenario is described below.

- At time $t = t_1$, task τ_3 executes *WAIT*(S_1), thus, locking semaphore S_1, and executes into its critical section.
- At time $t = t_2$, task τ_2 preempts task τ_3 and begins execution.
- At time $t = t_3$, task τ_2 attempts to lock semaphore S_2 by executing *WAIT*(S_2) but is prevented from doing so by the provisions of the priority ceiling protocol—task τ_2 is currently executing at its assigned priority, which does not exceed the priority ceiling of the currently locked semaphore S_1. Task τ_3 then resumes execution at the inherited priority of task τ_2. Task τ_2 is ceiling blocked, at this point, and not allowed to lock semaphore S_2.
- At time $t = t_4$, task τ_1 preempts task τ_3 and begins execution.
- At time $t = t_5$, task τ_1 attempts to execute *WAIT*(S_2). In accordance with the provisions of the priority ceiling protocol, task τ_1 is granted the lock on semaphore S_2 because its priority is higher than the priority ceiling of the currently locked semaphore S_2. Task τ_1 continues execution in its critical section.
- At time $t = t_6$, task τ_1 executes *SIGNAL*(S_2), thus, unlocking semaphore S_2, and exits its critical section.

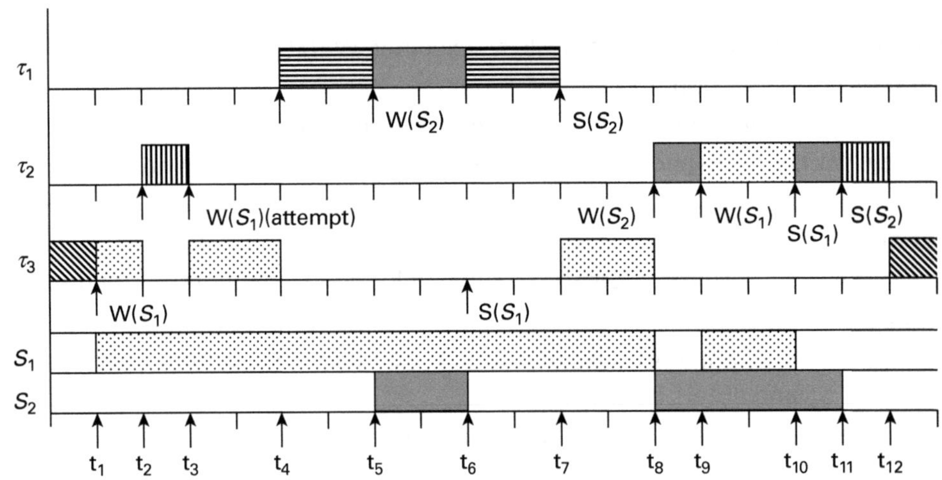

Figure 6.16 Prevention of Transitive Blocking by the Priority Ceiling Protocol (S_1 Priority Ceiling = Priority of Task τ_2 and S_2 Priority Ceiling = Priority of Task τ_1).

- At time $t = t_7$, task τ_1 completes. Since task τ_2 remains ceiling blocked, task τ_3 resumes execution at the inherited priority of task τ_1.
- At time $t = t_8$, task τ_3 executes *SIGNAL*(S_1), thus, unlocking semaphore S_1. Since there are no locked semaphores at this point, task τ_2 is no longer ceiling blocked. Task τ_3 reverts to its assigned priority and is preempted by task τ_2, which immediately locks semaphore S_2 and executes into its outer critical section.
- At time $t = t_9$, task τ_2 executes *WAIT*(S_1), thus, locking semaphore S_1, and executes into its inner critical section. Task τ_2 is permitted the lock on semaphore S_1 because even though its priority does not exceed the priority of currently locked semaphore S_2—task τ_2, itself, currently holds the lock on semaphore S_2.
- At time $t = t_{10}$, task τ_2 executes *SIGNAL*(S_1), thus, unlocking semaphore S_1, and exits its inner critical section.
- At time $t = t_{11}$, task τ_2 executes *SIGNAL*(S_2), thus, unlocking semaphore S_2, and exits its critical section.
- At time $t = t_{12}$, task τ_2 completes, permitting task τ_3 to resume execution.

It is the ceiling blocking of task τ_2 at time $t = t_3$, when it attempts to lock semaphore S_2, that prevents the transitive blocking of high priority task τ_1 that occurs in the scenario depicted in Figure 6.13 even with the basic priority inheritance protocol in effect.

6.10.4 Summary of Reliability Protocols

In dynamically scheduled systems, task synchronization applied through the use of blocking mechanisms such as semaphores can cause unbounded priority inversion. Priority inversion, on the other hand, prevents any kind of schedulability analysis because the possibility of long durations of priority inversion prevents the determination of worst-case blocking times for high priority tasks. The operational counterpart to this problem is that high priority tasks may miss their deadlines.

The problem of priority inversion can be addressed through the use of effective priority inheritance synchronization protocols. The basic priority inheritance protocol prevents unbounded priority inversion. The priority ceiling protocol prevents unbounded priority inversion and, at the same time, prevents chain blocking and transitive blocking, effects that may be present with the use of the basic priority inheritance protocol. The priority ceiling protocol also prevents deadlock, which can occur with the use of the basic priority inheritance protocol.

In summary, the priority ceiling synchronization protocol provides for the control of priority inversion in dynamic scheduling of dependent tasks that must synchronize. A task can be blocked by lower priority tasks for, at most, the duration of one enclosing critical section—there is no chain blocking—and there is no transitive blocking. In addition, use of the priority ceiling protocol prevents deadlock conditions from developing. The prevention of unbounded priority inversion—as is accomplished with either the basic priority inheritance protocol or with the priority ceiling protocol—is necessary before schedulability analysis techniques can be applied to task systems in which blocking effects are present.

CH 7

REAL TIME SYSTEM APPLICATION NOTES

This chapter will discuss several additional topics and techniques that are very useful while designing real time systems that are reliable in nature, i.e., meet deadlines and system specifications in a timely fashion. We will first discuss the non-preemptive static scheduling of periodic tasks on one processor—also known as a *cyclic executive*. The topic of handling aperiodic tasks, specifically, *sporadic* tasks, is, then, addressed in a formal fashion.

7.1 The Cyclic Executive Revisited

Many real time systems utilize a very simple form of scheduling known as a *cyclic executive*. It was first addressed in Chapter 4 and referred to in the periodic task scheduling section in Chapter 5. Such a system is designed for low cost and efficiency without the overhead of a RTOS and basically involves the non-preemptive static scheduling of periodic tasks on one processor. It is popular for the reasons mentioned above but has drawbacks as the system becomes more complex or is modified.

The term static task scheduling refers to the process of determining a fixed order of execution of the tasks in a task system, an order that repeats itself in a periodic manner as the system executes. The order never changes, and the periodicities are produced by timer-driven interrupts. In practice, there are several factors, such as jitter and the servicing of aperiodic tasks, that can cause modification of the fixed order. However, it is assumed for modeling purposes that the real time systems developer must address these issues in the design.

The software that responds to the timer-driven interrupts by scheduling tasks in the prescribed order is called a *cyclic executive*. In order to design a cyclic executive, it is first necessary to develop the periodic task schedule. The basic purpose of the task schedule is to provide the necessary execution time for each task while meeting the deadline and periodic requirements of the task. The design of a cyclic executive schedule can be difficult since the scheduling of periodic tasks with arbitrary release times and deadlines is a difficult problem. Two techniques are commonly applied.

- The task system can be forced into a configuration that eases the complexity of the task-scheduling problem. A common method is to force the tasks to have repetition rates that are

powers of two (2 Hz, 4 Hz, 8 Hz, 16 Hz, etc.). The scheduling of systems of tasks with this characteristic is trivial.
- Heuristic algorithms can be applied to scheduling problems.

It is common in the design of cyclic executive task schedules for some tasks to be divided into smaller subtasks for the purpose of task synchronization. Further division of the subtasks is often required to fit them into the cyclic schedule in a manner that ensures their deadline requirements are met. This kind of subdivision of subtasks requires arbitrary preemption. The result is often brittle in the sense that small design changes can cause a major redesign of the schedule. While later developments deal with the division of tasks into subtasks, the current discussion considers the scheduling of non-preemptive tasks.

It is important to recognize, as the discussion of periodic task scheduling and the design of cyclic executives proceeds, that the purpose of the schedule is not to optimize (minimize) the time of execution of a task system—as is the case in the scheduling of aperiodic tasks described in Chapter 4—but, rather, to fit the execution of the tasks into the available time. Figure 7.1 shows a simple cyclic executive schedule that will be used to illustrate this concept. The baseline system treated here is the single processor system, so the scheduling problem considered here is that of designing a cyclic schedule so that the tasks will successfully execute on a single processor. In Figure 7.1, p_1, p_2, and p_3 designate the periods of the three tasks τ_1, τ_2, and τ_3, respectively. The deadline of each of the tasks is the period of the task.

- A *cyclic schedule* is a schedule of interleaved subtasks that will enable the subtasks to execute within their deadlines and meet periodic constraints. Subtasks, once scheduled, execute to completion. These schedules characteristically consist of identifiable subunits called the major and minor cycles. The tasks of the simple example shown in Figure 7.1 have not been divided into subtasks. A subtask, in the sense used here, may be a task that has not been subdivided.

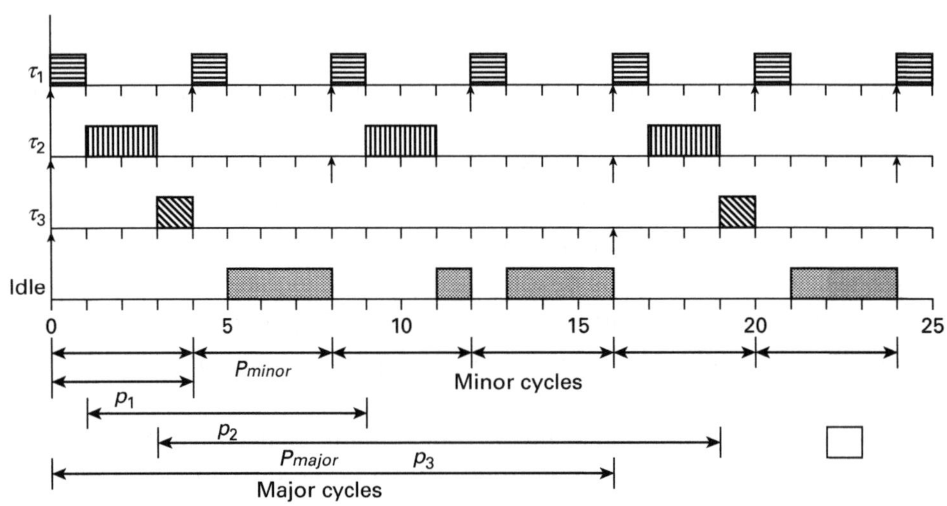

Figure 7.1 A Cyclic Schedule.

- The *major cycle* is a fixed cycle of subtask executions during which all tasks complete their execution. Each major cycle is identical to every other major cycle. The period of the major cycle will be designated by P_{major}.
- The major cycle is composed of a number of *minor cycles*. The period of each minor cycle is the same and is usually the period of the task with the highest periodic rate. It is the minor cycle that is enforced by timer-driven interrupts. The period of the minor cycle will be designated by P_{minor}. In the cyclic schedule illustrated in Figure 7.1, there are four minor cycles comprising the major cycle.
- The sequence of subtasks assigned to a minor cycle is called a *frame*. While each minor cycle has the same period, different subtasks are executed within different frames. Any particular frame repeats at the major cycle rate. There are four frames, one corresponding to each minor cycle. The first frame contains the execution of tasks τ_1, τ_2, and τ_3; the second frame contains the execution of task τ_1 followed by a period of idle time; the third frame consists of the execution of tasks τ_1 and τ_2, followed by a period of idle time; and the fourth frame, again, contains the execution of task τ_1 followed by a period of idle time. It is typical that cyclic scheduling produces times in the schedule during which no tasks are executing. These times are shown in Figure 7.1 as processor idle time. Actually, processors are never idle in the strict sense so that this time is normally used to perform background processing, non-time-critical processing such as diagnostic checks and other tasks termed *general housekeeping*.
- Because of effects like jitter or the execution of aperiodic tasks, it is not uncommon for a frame's minor cycle to occasionally terminate before all the tasks and/or subtasks comprising the frame have completed execution. This situation is called *frame overrun* and represents a potentially serious problem.
- In many real time systems, there are various *modes* of operation. For example, an automobile engine might be in start mode or in idle mode; a radar might be in search mode or track mode. The effect of these differing modes of operation on the execution of the real time software is usually a change in the major cycle. Typically, tasks may be deleted and/or additional tasks may be added, forcing the software developer to consider the dynamic effect such *mode changes* have on the execution of the software. Mode changing is discussed later.

7.1.1 Cyclic Scheduling of Independent, Non-Preemptable Tasks

The task systems to be considered here consist of tasks that are independent—they have no precedence or synchronization requirements—and are not preemptable. That is, the tasks cannot be divided into subtasks for purposes of scheduling. Under these conditions, the values of the major and minor cycle periods are subject to a set of requirements that result from relationships among the various task parameters.

1. The minor cycle period P_{minor} must be no larger than the period of the highest rate task. This requirement arises from the fact that the periodicities of all the tasks are derived from the minor cycle period. In essence, P_{minor} is the period of the sequence of timer-driven interrupts. Timing signals (timer-driven interrupts) are available only at the beginning of a minor cycle. Tasks executed within a minor cycle are sequenced rather than scheduled.

2. The minor cycle period P_{minor} must be at least as large as the largest execution time of any of the tasks. This requirement follows from the fact that each of these tasks must execute to completion within a minor cycle.
3. The major cycle period P_{major} is, by definition, the least common multiple (lcm) of the individual task periods $\{p_1, p_2, \cdots, p_n\}$. That is, the pattern of the release times of the individual tasks repeats with a period of $lcm\{p_1, p_2, \cdots, p_N\}$. The minor cycle period P_{minor} must evenly divide into P_{major}, the major cycle period. This requirement arises from the fact that there are an integer number of minor cycles in a major cycle.
4. If the minor cycle period P_{minor} must be selected so that each task has the flexibility of executing for the duration of a complete minor cycle, then it is necessary to have a complete frame between the release time of a task and the associated deadline. This requirement can be stated as Equation 7.1.1

$$2\,P_{minor} - gcd\left\{P_{minor}, p_n\right\} \leq d_n \qquad \text{(EQ 7.1.1)}$$

Here, gcd is the greatest common divisor, the largest integer that will evenly divide both P_{minor} and p_n. Such flexibility is implied by requirement (2) but may not be necessary in specific cases.

Requirement (4) above results from the characteristic that unless p_n—the period of task τ_n—is harmonically related to the minor cycle period P_{minor}, its release times over the extent of a major cycle will wander over the entire length of the minor cycle. The worst case of this wander results when the release time occurs as soon as possible after the beginning of a minor cycle. That is, since the task is scheduled on the basis of the start of a minor cycle—the occurrence of a timer-driven interrupt—the maximum delay results when the release time occurs immediately following the beginning of a minor cycle since the task cannot be scheduled until after the next minor cycle begins. This effect is illustrated in Figure 7.2, which depicts a sequence of minor cycles and two successive release times of task τ_n—the k-th and the $(k + 1)$th. Between these two release times, some number of minor cycles take place. The period p_n and the deadline d_n of task τ_n are identified in the figure. The minimum non-zero time that can separate the beginning of a minor cycle and the release time of task τ_n is $gcd(P_{minor} > p_n)$ as indicated in the figure. The result of Equation 7.1.1 then follows.

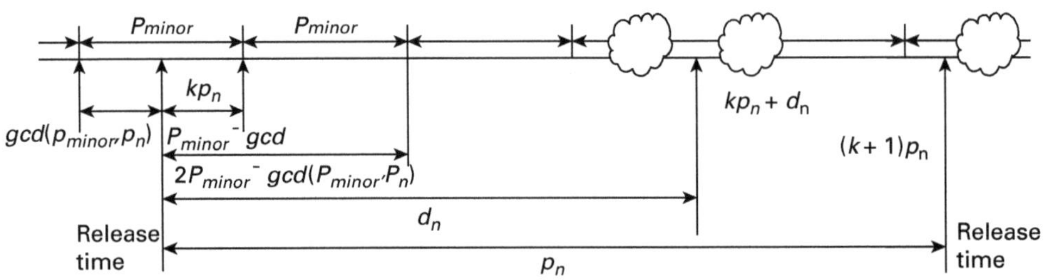

Figure 7.2 The Worst-Case Phasing of Release Times.

These concepts can be illustrated through the use of some simple examples that show the variation in the phasing of task release times over a complete major cycle.

Example 7.1.1: An example of how task release times vary in phase over a major cycle.

Figure 7.3 depicts the phasing between the release times of a task with a period of 12 and a minor cycle with a period of 10. The task release time and the minor cycle are in phase at time $t = 0$ and at time $t = 60$. (The value 60 is $lcm(10, 12)$, the major cycle period). In this case, $gcd(10, 12)$ is 2, and the worst-case delay occurs at time $t = 12$ where the task is released two time units after the beginning of the minor cycle at time $t = 10$. That is, the earliest the task can be scheduled is $t = 20$, the beginning of the next minor cycle. The reader should note the wander of the release times of the task within the minor cycles over the expanse of the major cycle.

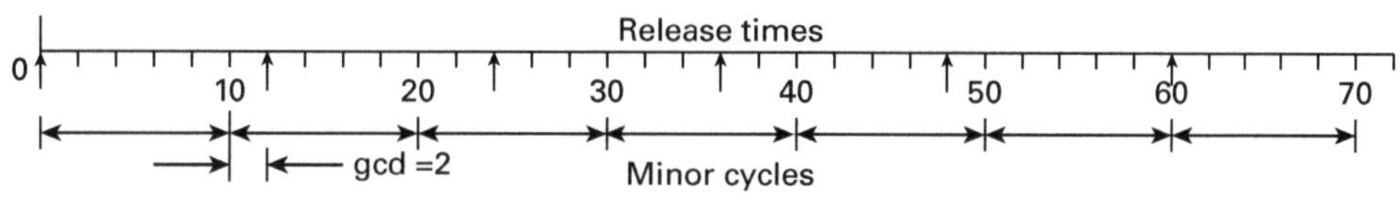

Figure 7.3 Phasing of a Task of Period 12 and a Minor Cycle of Period 10.

Example 7.1.2: An example with relatively prime periods.

A second example is shown in Figure 7.4, in which the phasing is depicted for a task of period 12 and a minor cycle of period 10. In this case, the two periods are relatively prime. Thus, $gcd(5, 12)$ is 1, and the worst-case phasing occurs when the task is released at time $t = 36$. The task cannot be scheduled until after the beginning of the following minor cycle at time $t = 40$. The worst-case delay in this case is four time units. Two relatively prime integers have no common divider other than 1.

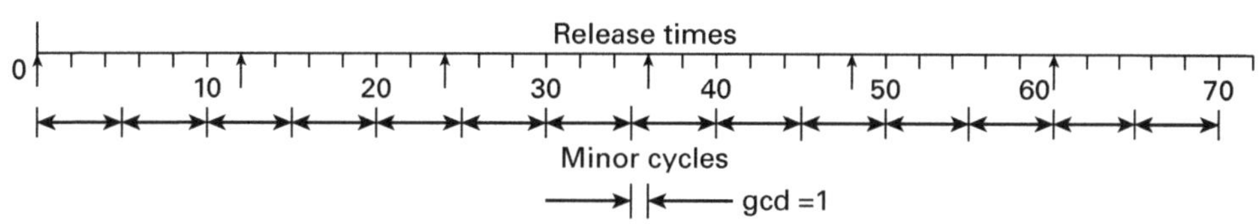

Figure 7.4 Phasing of a Task of Period 12 and a Minor Cycle of Period 5.

The requirements outlined lead to some interesting conclusions.

- It is *not sufficient*, although it is necessary, that

$$P_{minor} \leq min\{p_1, p_2, \cdots, p_n\} \qquad \text{(EQ 7.1.2)}$$

  This implies that the smallest period may not be a suitable selection for P_{minor}.
- It is *not necessary*, although it is sufficient, that

$$P_{minor} \leq gcd\{p_1, p_2, \cdots, p_n\} \qquad \text{(EQ 7.1.3)}$$

- Harmonic relationships among the task periods greatly simplify the scheduling problem. In particular, this kind of relationship eliminates the phasing problems described in conjunction with Figures 7.3 and 7.4, and it tends to produce small major cycle periods. As can be surmised, harmonic relationships among task periods also produce higher processor utilization because they eliminate the requirement for including processor idle time when developing cyclic schedules.

7.1.2 Task Execution Times and Processor Utilization

In the selection of major and minor cycle periods, the only effect of execution times is the requirement that the minor cycle period be at least as long as the execution time of the longest subtask. Obviously, the execution times of the tasks in a task system are critical to the development of a cyclic schedule. Perhaps even more important is the effect of task execution times on the answer to the YES/NO question, is it possible to schedule a particular task system on a single processor? One rather obvious limitation is that the total execution time over a major cycle can be no larger than the major cycle period P_{major}. This requirement can be stated as the inequality

$$U = \sum_{1}^{N} \frac{e_n}{p_n} = \sum_{1}^{N} U_n \leq 1 \qquad \text{(EQ 7.1.3)}$$

where U is called the *processor utilization* and $U_n = \frac{e_n}{p_n}$ is the utilization of task τ_n. This form of the requirement results from the fact that the major cycle period is the least common multiple of the individual task periods, which can be expressed in the following manner.

$$P_{major} = lcm\{p_1, p_2, \cdots, p_n\} = \frac{p_1\ p_2 \cdots p_n}{k} \qquad \text{(EQ 7.1.4)}$$

where k is some integer greater than or equal to 1. When $k = 1$, the task periods are all relatively prime, and the P_{major} is the product of the individual task periods. This situation gives some insight into the complexity of the general periodic task-scheduling problem. The number of times a given task τ_n executes in a major cycle is

$$\frac{P_{major}}{p_n} = \frac{1}{k}\{p_1, p_2, \cdots, p_{n-1}, p_{n+1}, \cdots, p_n) \qquad \text{(EQ 7.1.5)}$$

Then, the execution time required by τ_n over a complete major cycle is its execution time multiplied by the number of times that it executes, as given by Equation 7.1.6.

$$\frac{e_n}{k}\{p_1, p_2, \cdots, p_{n-1}, p_{n+1}, \cdots, p_n) \qquad \text{(EQ 7.1.6)}$$

The total execution time over a complete major cycle is obtained by summing the execution times of all the tasks, and this result must not exceed the major cycle period as given by Equation 7.1.4.

$$\frac{1}{k}\sum_{1}^{N} \frac{e_n}{k}\{p_1, p_2, \cdots, p_{n-1}, p_{n+1}, \cdots, p_n) \leq \frac{p_1\ p_2 \cdots p_n}{k} \qquad \text{(EQ 7.1.7)}$$

Canceling terms in Equation 7.1.7 results in the inequality of Equation 7.1.3, which requires that the processor utilization—the sum of the individual task utilizations—must be less than one. The processor utilization of a task system will play an important role in the schedulability analysis to follow.

Example 7.1.3: Major and minor cycle selection.

The concepts outlined above can be illustrated by the following example. The problem is expressed as N, the number of tasks, and the values of p_n, the task periods; d_n, the task deadline; and e_n, the execution time of the task.

$$N = number\ of\ tasks = 3$$

Period	Deadline	Execution Time
$p_1 = 4$	$d_1 = 4$	$e_1 = 1$
$p_2 = 6$	$d_2 = 6$	$e_2 = 1$
$p_3 = 10$	$d_3 = 10$	$e_3 = 2$

The processor utilization is determined to be $U = \frac{1}{4} + \frac{1}{6} + \frac{2}{10} = 0.616$.

It is important to realize just what this result implies on the basis of Equation 7.1.3. It does not show that the task system can be scheduled—it simply fails to show that the task system cannot be scheduled. The only way to show that the task system can be scheduled is to build a valid schedule, which—if it exists—will have a major cycle period given by Equation 7.1.4:

$$P_{major} = lcm\{p_1, p_2, p_3\} = 60$$

Now the four requirements of cyclic scheduling will be applied. The first requirement is:

$$P_{minor} \leq min\{p_1, p_2, p_3\} = \min\{4, 6, 10\}$$

This requirement limits the selection of the minor cycle period P_{minor} to the values 1, 2, 3, and 4. The second requirement is:

$$P_{minor} \geq max\{e_n\} = \max\{1, 1, 2\} = 2$$

Values of all times are in integral numbers of time units. This is reasonable since the only possible division of time is related to the period of the timer-driven interrupts.

This requirement further limits selection to the values 2, 3, and 4. The third requirement is that P_{minor} evenly divides P_{major}—that is P_{minor} evenly divide into 60. No further restrictions result from this requirement.

The fourth requirement is related to the phasing between task release times and the minor cycle and is given by Equation 7.1.1:

$$2\,P_{minor} - gcd\left\{P_{minor},\ p_n\right\} \leq d_n$$

Applying this relation to the three remaining candidate values of P_{minor} yields the following results.

$P_{minor} = 2$	$4 - gcd(2, 4) = 2 < 4$	ACCEPT
	$4 - gcd(2, 6) = 2 < 6$	ACCEPT
	$4 - gcd(2, 10) = 2 < 10$	ACCEPT
$P_{minor} = 3$	$6 - gcd(3, 4) = 5 > 4$	FAIL
	$6 - gcd(3, 6) = 3 < 6$	ACCEPT
	$6 - gcd(3, 10) = 5 < 10$	ACCEPT
$P_{minor} = 4$	$8 - gcd(4, 4) = 4 = 4$	ACCEPT
	$8 - gcd(4,6) = 6 = 6$	ACCEPT
	$8 - gcd(4,10) = 6 < 10$	ACCEPT

This requirement rules out the value 3 as a candidate minor cycle period.

The fact that the candidate P_{minor} values 2 and 4 meet the set of requirements does not necessarily imply that a valid schedule exists. The execution times of the tasks now determine if there is a valid schedule, and the only way in which the question can be answered is to actually build a schedule. There are some heuristic algorithms that aid in schedule building, but they are not guaranteed to produce a valid schedule even if one exists. One of these heuristic algorithms calls for the tasks to be scheduled in a priority manner, with the priorities assigned in decreasing order of task periods. Figure 7.5 shows the results of applying this algorithm to the current problem, using a minor cycle period of 2. This priority assignment is termed rate-monotonic and has been covered in Chapter 5.

It can be seen that all tasks meet their deadlines—which, of course, answers the YES/NO question of whether or not a valid schedule exists. In the development of this schedule, it is important to recall the requirement—discussed in relation to the situation depicted in Figure 7.2—that a task cannot be

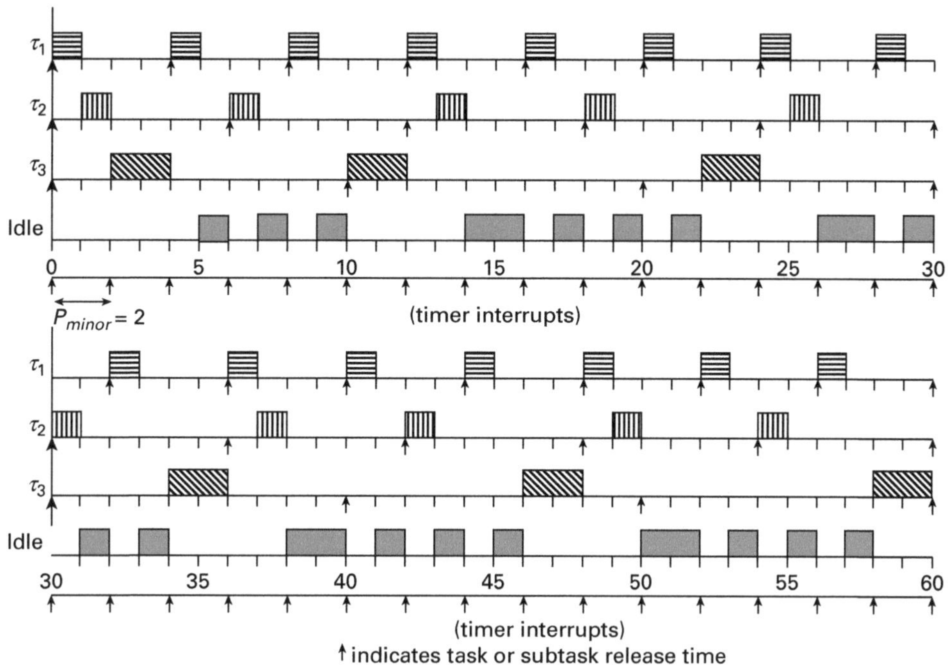

Figure 7.5 Cyclic Schedule with Minor Cycle Period = 2.

scheduled until the beginning of the first minor cycle following its release time. This requirement arises from the fact that the highest rate task is scheduled by the timer-driven interrupt sequence while all other tasks are sequenced from the execution of that task. If not for the requirement above, a task could be scheduled by sequencing prior to its release time if one or more of the tasks preceding it in a frame complete early. The task maybe scheduled immediately if its release time is concurrent with the beginning of a minor cycle.

Figure 7.6 shows the schedule derived using a minor cycle period of 4. These task schedules are not unique since there are a myriad of ways in which the task executions could be moved without violating any deadlines.

The value three was eliminated as a candidate minor cycle period by the fourth requirement for cyclic scheduling—expressed as Equation 7.1.1. This requirement exists to ensure the presence of at least one complete frame between the release time of a task and the corresponding deadline. The necessity of having at least one complete frame between the release time and the deadline of a task is related to the second requirement of cyclic scheduling—that the minor cycle period be at least as large as the execution time of any of the tasks. The implication of this requirement is that the execution time of *any* task may be as large as the minor cycle period. The fact that the tasks and/or subtasks of a cyclic schedule—once execution begins—must run to completion leads directly to the necessity of having at least one complete frame between any release time and the corresponding deadline.

The fourth requirement for cyclic scheduling—expressed as Equation 7.1.1—*does not consider the execution times of the tasks.* Rather, it is based on a worst-case situation existing for each task—that its execution time is the maximum possible, the duration of the minor cycle period. The point to be made is that requirement four may not be applicable when applied to particular tasks within a task system, with the result that some candidate minor cycle periods may be eliminated from consideration unnecessarily. This is the case with the candidate value of three, which was eliminated on the basis that

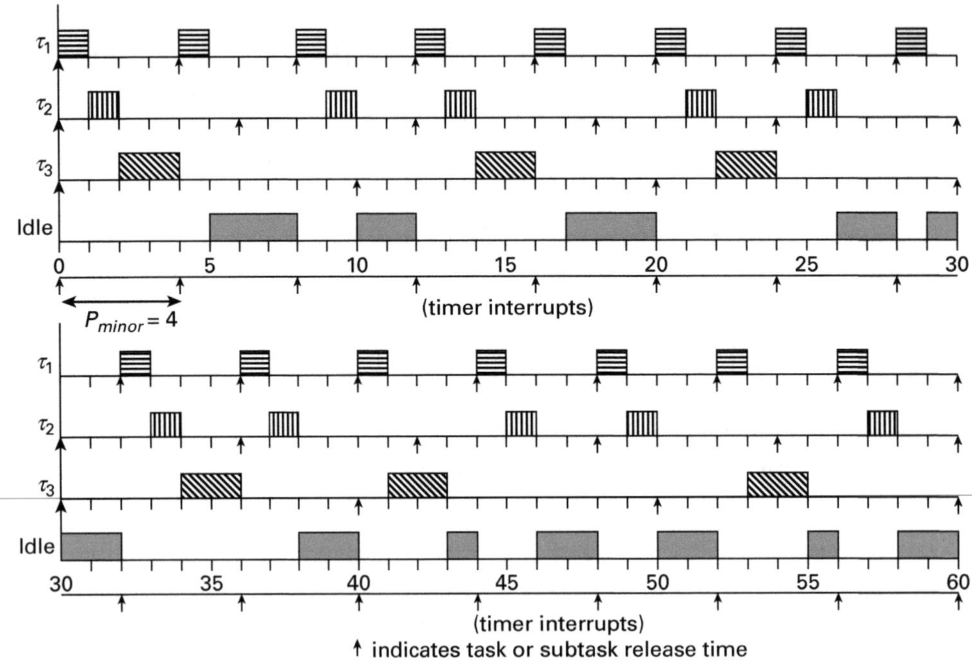

Figure 7.6 Cyclic Schedule with Minor Cycle Period = 4.

it violated requirement four with respect task τ_1. This situation is illustrated by the scheduling diagram of Figure 7.7, which contains a valid schedule for the task system of Example 7.1.3, obtained using a minor cycle period of three. The reason the value three failed the condition of Equation 4.2 can be seen by examining the time line of task τ_1. The worst-case phasing between the minor cycle and the release

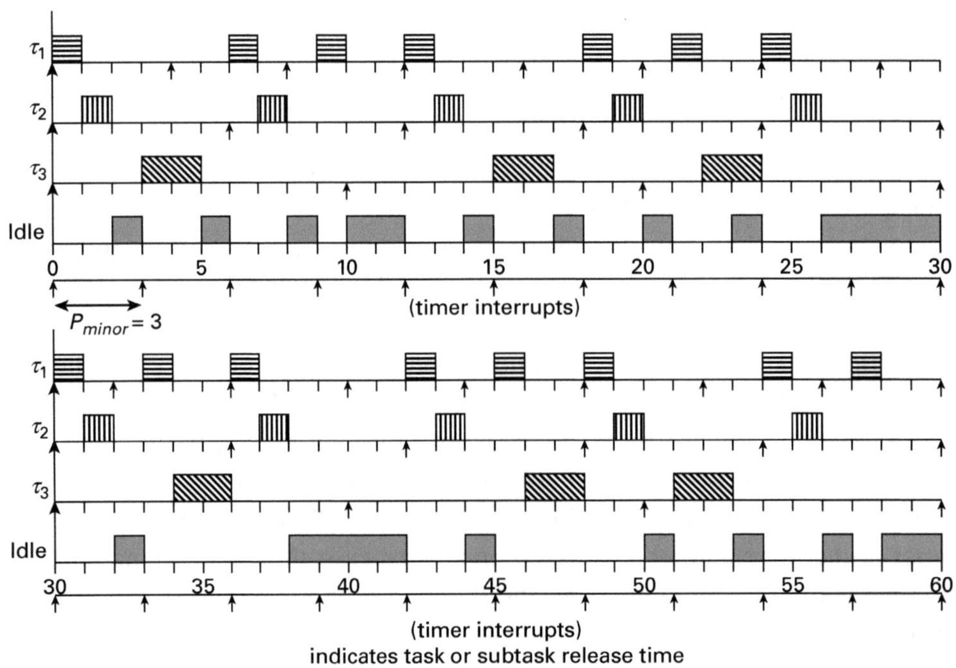

Figure 7.7 Cyclic Schedule with Minor Cycle Period = 3.

times of task τ_1 occurs for the first time at time $t = 4$ when task τ_1 is released one time unit after the beginning of a minor cycle; note that the value of $gcd(P_{minor}, p_1)$ is $gcd(3, 4) = 1$.

In this particular case, the execution time of task τ_1 is one time unit, which does not present a problem. In the general case, however, the only requirement placed on the relationship between the minor cycle period and the execution time of task τ_1 is that of the second requirement of cyclic scheduling—that P_{minor} be *no larger than* e_1, the execution time of task τ_1. It is easy to see from the situation existing at time $t = 4$ in the diagram of Figure 7.7 that the execution time of task τ_1 cannot exceed the value two. This limitation violates the general condition that any task period may be as large as the minor cycle period. The point to be made here is, simply, that requirement four of cyclic scheduling is a general worst-case requirement that may require more detailed evaluation in certain cases.

Requirement four does not involve the values of the task execution times and, thus, does not directly consider the schedulability of the task system. This can be illustrated by noting that a value of three for the execution time of task τ_1 in this example would result in a processor utilization value of 1.117 and a clearly non-schedulable task system.

The task-scheduling process described above deals with independent, non-preemptable tasks. Independent tasks have no precedence or synchronization constraints that must be met by the scheduling process. As might be expected, task dependencies such as precedence constraints and synchronization requirements considerably complicate the problem of task scheduling. The effects of task dependencies such as these are addressed in the following section.

7.1.3 The Effects of Task Precedence and Synchronization Constraints

Precedence constraints complicate the scheduling process by imposing restrictions on the way tasks can be interleaved for execution. Synchronization constraints, on the other hand, require the division of tasks into subtasks and the scheduling of these subtasks in a manner that implements the required synchronization. The basic concept can be illustrated by the simple task schedules depicted in Figure 7.8, in which the two tasks τ_1 and τ_2 are divided into the sets of subtasks $\{\tau_{11}, \tau_{12}, \tau_{13},\}$ and $\{\tau_{21}, \tau_{22}, \tau_{23},\}$, respectively, in order to meet synchronization requirements relating to the points within tasks τ_1 and τ_2 at which the divisions are made.

The precedence diagram implies, for example, that subtask τ_{12} cannot begin execution until subtasks τ_{11} and τ_{21} have completed. Similarly, task τ_{22} cannot begin execution until subtasks τ_{21} and τ_{12} have completed, and so on. Without these synchronization requirements—that is, if tasks τ_1 and τ_2 were independent—then the schedule could be implemented simply by executing task τ_1 followed by task τ_2 or vice versa. With the division of tasks τ_1 and τ_2 into subtasks as shown, the synchronization constraint can be satisfied by interleaving execution of the subtasks—as shown in Figure 7.8—in a manner that implements the required synchronization.

Another kind of synchronization requirement is the exclusion relation, in which one subtask must not be scheduled between the execution of two other subtasks. Such situations commonly arise because of exclusionary access to some kind of shared resource or data structure. The precedence relation of Figure 7.8, modified to include the exclusion of subtask τ_{21} from the sequencing of subtasks τ_{12} and τ_{13}, is depicted in Figure 7.9, together with a schedule that implements the exclusion.

The division of tasks into subtasks in order to implement synchronization constraints further complicates the already complex problem of task scheduling. The problem of scheduling *independent* tasks on one processor can be difficult, and the addition of synchronization requirements merely adds to the difficulty of developing a schedule—*if a valid schedule exists*—by heuristic means.

The point to be made is that the task-scheduling problem—an important step in the development of real time software—is *difficult*. The general problem can be theoretically hard, implying that there is no general algorithm for producing a valid schedule for a given task system. Even worse, there is no way of answering the YES/NO question, is it possible to construct a valid schedule? The only way to answer this question is to actually build a valid schedule. The difficulty with this approach is that the only way in which one can arrive at a negative answer to the YES/NO question is to try all possible schedules, a characteristic of theoretically hard problems.

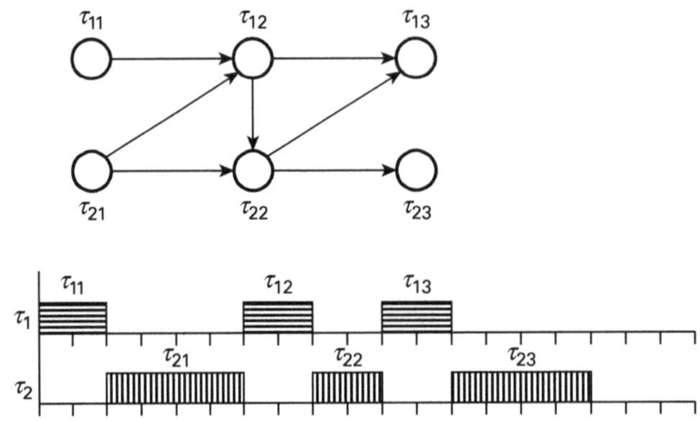

Figure 7.8 Precedence Graph Specifying Precedence Relations Among Subtasks.

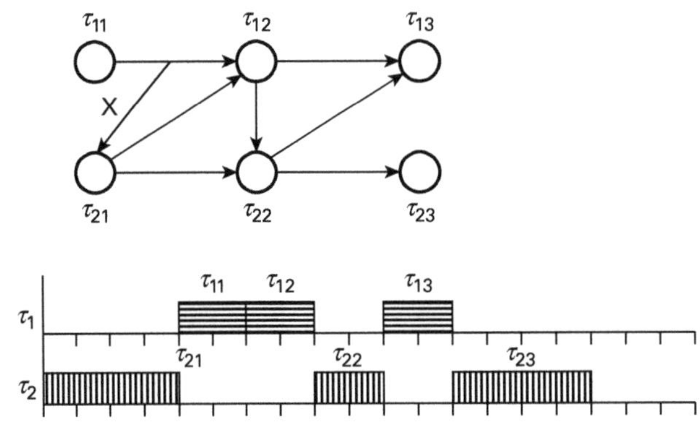

Figure 7.9 Modified Precedence Graph Specifying Precedence Relations Among Subtasks.

The prudent approach is to recognize that the general problem is difficult and use some kind of heuristic algorithm. This approach was discussed in detail in Chapter 5, in which scheduling algorithms such as rate-monotonic scheduling, earliest deadline scheduling, and deadline-monotonic scheduling were described.

7.1.4 Providing for Aperiodic Task Execution

In addition to the periodic tasks that execute synchronously within the cyclic schedule, most real time systems have asynchronous processing requirements in the form of aperiodic tasks. Usually, these aperiodic tasks are few in number, require a relatively small amount of computation, and exhibit a predictable minimum interarrival time. In the simplest of situations, these aperiodic tasks are handled as interrupts, using the hardware interrupt processing capabilities of the processor. Interrupt processing occurs at the highest level of priority within the processor and, therefore, can severely disrupt the periodic schedule. Interrupts, of course, can be masked to protect periodic tasks from disruption, but this approach becomes unwieldy with large numbers of tasks and deadline sensitive aperiodic tasks.

A second approach to aperiodic task scheduling is to relegate the execution of aperiodic tasks to background processing. This technique can be implemented by having the interrupts masked during the execution of the periodic tasks, thus, postponing interrupt processing until the background processing becomes active. This approach suffers from what may be unacceptable delays in the execution of the aperiodic tasks. A method is needed for handling aperiodic tasks in a manner that provides timely response to the aperiodic request without disrupting the periodic schedule.

If an aperiodic task can be characterized by

- an execution time e_a,
- a deadline d_a, and
- a minimum interarrival time, *mit* , that specifies the minimum time that may elapse between one release time of the task and the next,

the task is called a *sporadic task*. The interarrival times are asynchronous—essentially random—but the physical process responsible for the production of the aperiodic request for service may be limited in how soon after one request it can generate the following request.

Figure 7.10 illustrates the concept of a task with a minimum interarrival time *mit* and deadline d_a. When the aperiodic request arrives—that is, when the aperiodic task is *released*—it may not be serviced immediately due to other pressing needs related to periodic execution or, perhaps due to other aperiodic processing requirements. The aperiodic task has a deadline, as shown in the diagram of Figure 7.10, which requires that the aperiodic task must complete within d_a time units of its release. The next release time cannot occur until *mit* time units after the current release time. The diagram of Figure 7.10 illustrates a possible scenario involving two release times of a sporadic task. They are separated by a time interval greater than the minimum interarrival time *mit*.

Sporadic tasks are often accommodated in a cyclic task schedule by dedicating a *periodic* task to provide the needed execution time. The periodic task—called a *polling task*—is a member of the periodic task system to which the scheduling techniques of the previous sections are applied. It, therefore, executes in the same periodic manner as the other periodic tasks. When the polling task executes, it initiates a query to determine if the aperiodic task has been released, in which case, it executes the aperiodic task. If the response to the query is no, the polling task terminates, and execution of the periodic tasks—or, more likely, the execution of background processing—resumes.

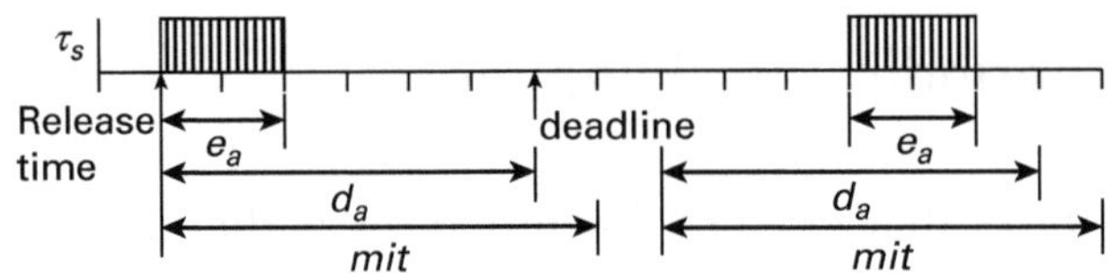

Figure 7.10 A Sporadic Task.

The period, execution time, and deadline of a polling task are selected on the basis of the characteristics of the aperiodic task it is to serve. If the aperiodic task τ_a has execution time e_a and deadline d_a, then the polling task is designed with the following properties.

- The execution time of polling task τ_p is selected to be $e_p = e_a$.
- The deadline of the polling task is selected to be $d_p = e_a$.
- The period of the polling task is selected to be $p_p = \min\{(d_a - e_a + 1), mit\}$

The use of a polling task is illustrated by the scenario depicted in Figure 7.11, which shows the execution of the periodic polling task τ_p overlaid with the execution of an aperiodic task τ_a. This scenario depicts three releases of the aperiodic task τ_a occurring over a time span in which the polling task τ_p is released five times. Even though the diagram of Figure 7.11 shows two tasks, there is a single task execution. The execution time of the polling task is allocated either to the aperiodic task or to background or other processing as specified by the cyclic executive. The execution of the polling task is described below.

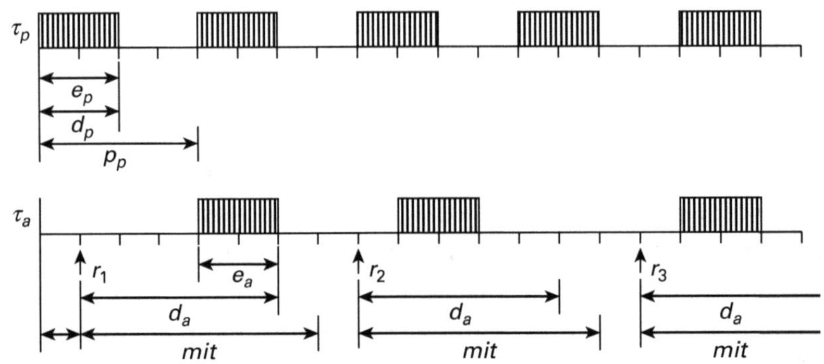

Figure 7.11 The Use of a Polling Task.

At the time of the first release of the polling task, no aperiodic request has been received. Hence, the polling task will terminate, and background processing or some other task will begin and will consume the execution time of the polling task.

- After the polling task has terminated, the aperiodic task is released—but it must now await execution until the polling task is released again.
- At the second release of the polling task, there is an aperiodic request pending, causing the polling task to immediately execute the aperiodic task.
- Before the polling task is released for the third time, a second aperiodic request arrives so that the third release of the polling again results in the execution of the aperiodic task.

- At the fourth release of the polling task, there is no aperiodic request pending, and background or other processing is begun and consumes the execution time of the polling task.
- At the fifth release of the polling task, there is, again, an aperiodic release pending.

Since the polling task is part of the periodic task system for which the cyclic schedule is developed, the execution of an instance of the aperiodic task does not affect the cyclic schedule. And, with the period of the polling task selected as the minimum of $\{(d_a - e_a + 1), mit\}$, the aperiodic task request does not suffer large delays.

7.1.5 Mode Changes in Cyclic Systems

The concept of mode changes in cyclic executive software as a consequence of differing operational modes of the embedding system has been previously mentioned. There may be other reasons, as well, for initiating a mode change, such as some kind of throughput problem sensed by the system software. Whatever the cause, the usual effect of operational mode changes on the software is a modification of the major cycle. The effect of a mode change can range from a modified set of frame assignments to a completely different cyclic structure. Tasks may be added to the schedule, tasks may be deleted from the schedule, resource allocation to the tasks might be modified, or a combination of these processes, together with others not mentioned, may be required.

A basic design decision is that concerning at just what point a mode change—a modification of the major cycle—will be implemented. The choices are listed below.

- Implement the mode change immediately upon receiving indication of the requirement of a mode change. This choice will interrupt the currently executing task and, therefore, can present problems with data integrity.
- Implement the mode change upon completion of the currently executing frame—that is, at the end of a minor cycle. This choice may avoid some data integrity problems, but possible problems involving sequencing or other task dependencies may remain.
- Implement the mode change upon the completion of the currently executing major cycle. This choice has appeal because the major cycle is the basic periodicity of the entire system. That is, each major cycle is identical to every other major cycle, which tends to ensure that the function performed by a given task or set of tasks has completed one cycle and is ready to begin the next.

The design of the particular mode-changing method to be used must consider the integrity of data, the effect on non-periodic processing such as aperiodic task servicing and background processing, and hardware interface requirements that may change with the addition of new tasks or the modification of existing tasks.

7.2 Sporadic Task Scheduling

The scheduling of periodic task systems must often account for the presence of asynchronous processing requirements in the form of aperiodic tasks. When dynamic scheduling is used, the schedule designer has added flexibility that can be effectively used to provide better service to the aperiodic tasks without jeopardizing the integrity of the periodic schedule. Aperiodic tasks are released randomly—usually in response to an external stimulus such as action by an operator, the detection of some event, or the availability of data. In some situations, the aperiodic task may satisfy the requirement of a sporadic task, while in other situations, the aperiodic task may have other characteristics. The deadline requirements of aperiodic tasks may also vary, with some aperiodic tasks having hard deadlines—deadlines that must be met under all conditions—and others having soft deadlines that require only that they be met in an average manner. In general, three types of aperiodic tasks tend to occur.

- *Sporadic* tasks are those hard deadline aperiodic tasks for which the response time is short, there is a known upper limit on execution time, and task releases cannot occur closer together than some *minimum interarrival time*.
- *Bursty* tasks are those hard deadline aperiodic tasks that have burst characteristics—that is, task releases may tend to occur close together at certain points in the operation of the system. The response time is short, and there is an upper limit on execution time.
- *Unbounded* tasks are those soft deadline tasks for which average characteristics are known but whose interarrival times and execution times are not bounded.

The arrival times are described by some probability distribution characterized by parameters that correspond to ensemble averages. Actual occurrences (samples of the distribution) may be the result of samples at the extreme ends of the distribution and, thus, are unbounded. The Poisson arrival process is a good example (tasks arrival times may be characterized using the random Poisson distribution model).

Several methods for scheduling aperiodic tasks can be utilized. Obviously, the precise method to be used in any particular situation will vary depending on the system requirements and the characteristics of the aperiodic task itself. Those methods to be discussed are listed below.

- The use of a hardware interrupt and an associated interrupt handler to provide for the execution of the aperiodic task.
- The use of background processing to provide for the execution of the aperiodic task.
- The use of a periodically scheduled polling task.
- The use of one of several algorithms known as bandwidth preserving algorithms.

7.2.1 Interrupt Handling

The use of interrupt handlers is the most direct way of providing execution time to aperiodic tasks, and it holds the greatest potential for disruption of the periodic task schedule. Since external events requiring the release of aperiodic tasks are often announced by the occurrence of interrupts, this method is easy to implement. There are two basic approaches that may be taken.

- The entire aperiodic task may be executed as the interrupt handler at the interrupt priority level of the processor. This approach produces the shortest response time for the aperiodic task, but it also uses relatively large amounts of high priority processing time that may be required for meeting the periodic requirements of the periodic task system. While this approach can easily destroy the somewhat delicate schedule of a cyclic executive, it may be practical for some limited form of dynamically scheduled system in which the flexibility of the dynamic schedule permits adjustment to the scheduling of periodic tasks.

- Processing at the interrupt priority level of the processor may be limited to the time-critical portion of the aperiodic task and the remainder of the task completed at a clock level priority commensurate with the requirements of the periodic task system. This approach is compatible with those situations in which the interrupt announces that data of some sort is required or available. The interrupt can be handled very quickly at the interrupt level priority and consists of any data transfer required and of handshaking with the external device. Processing of the data would then be accomplished at clock level priority. In general, priority levels are interrupt level or clock level. Interrupt level priorities are those priorities levels provided by the processor for interrupt handlers. Clock level priorities are those priorities assigned to tasks by the real time multitasking executive.

Either of these approaches is subject to schedulability analysis using utilization bounds based on worst-case conditions if the aperiodic task is bounded—that is, if the task has limited execution time, and releases occur no closer together than some minimum interarrival time. The worst-case scenario occurs when the interrupt handler, which executes at a priority higher than any periodic task, is released at a rate corresponding to the minimum interarrival time, in which case, it can be treated as a high priority periodic task with that period.

In the case of an unbounded task, this approach is not feasible if the periodic schedule is to be preserved under all conditions since the unbounded aperiodic task, by definition, may experience periods of very rapid releases or execution times that are very long. This characteristic is the reason unbounded aperiodic tasks must have soft deadline requirements, in that while some deadline violations will occur, on the average, the deadlines will be met.

When the actual processing of data associated with an aperiodic task release has no particular deadline requirement, it may be relegated to background processing. Normally, background processing is allocated to the execution of non-critical tasks such as diagnostics or other forms of housekeeping duties, but there may be situations in which certain aperiodic processing requirements can be accommodated as a background task. In effect, using background processing for execution of an aperiodic task in a dynamically scheduled task system is an application of the method described above, that of using a clock level priority task to service an aperiodic task. In a cyclic executive system, this technique can be particularly useful since processing in the background does not interfere with the cyclic schedule.

Example 7.2.1: The use of background processing to service aperiodic tasks.

The use of background processing to execute an aperiodic task is illustrated using the task system specified below. Since deadlines equal periods, rate-monotonic priority assignment will be applied.

$N = number\ of\ tasks = 3$

M = number of processors = 1

Period	Deadline	Execution Time
$p_1 = 10$	$d_1 = 10$	$e_1 = 2$
$p_2 = 12$	$d_2 = 12$	$e_2 = 4$
$p_3 = 22$	$d_3 = 22$	$e_3 = 6$
τ_a *aperiodic*		$e_4 = 2$

The processor utilization is determined to be

$$U = \frac{2}{10} + \frac{4}{12} + \frac{6}{22} = 0.806 \leq 1.$$

The utilization bound for three tasks is 0.780, so the utilization bound is inconclusive. Nevertheless, the task periodic system is schedulable, as indicated by the schedule depicted in Figure 7.12. Also shown in this schedule is a release of the aperiodic task τ_a occurring at time $t = 4$. The occurrence of the task release might have been announced by an interrupt that was handled at an interrupt level priority at the time of occurrence in a negligibly short time. The point to be made by this example is the excessive delay that can be produced if the aperiodic task is released at a time when the task will have a considerable wait for the next available background processing time.

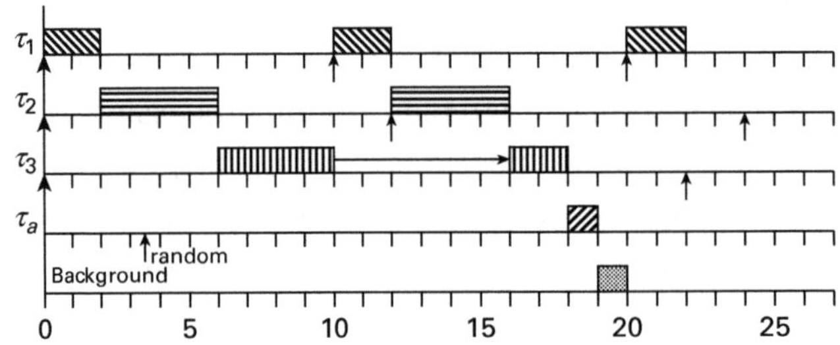

Figure 7.12 Background Processing of an Aperiodic Task.

7.2.2 Using a Polling Task

The use of polling tasks to provide for execution time in statically scheduled systems was considered in the previous section. The treatment here considers the use of a polling server task in a dynamically scheduled, priority-driven system. In this situation, the polling server task is usually given a relatively high priority to ensure timely response to the aperiodic task.

Example 7.2.2: Use of a polling task to service aperiodic tasks.

An example of a periodic task system that includes a polling server task is specified below. Rate-monotonic priority assignment will be used since it is an optimal static priority assignment algorithm when deadlines equal to periods.

$N = number\ of\ periodic\ tasks = 3$, *including the polling server task*

M = number of processors = 1

Period	Deadline	Execution Time
$p_1 = 8$	$d_1 = 8$	$e_1 = 2$
$p_2 = 12$	$d_2 = 12$	$e_2 = 4$
$p_s = 22$	$d_s = 6$	$e_s = 2$

The polling server task τ_s has been selected to have a period of 6 and an execution time of 2, based upon the particular characteristics of the aperiodic task(s) it is to serve. The prior discussion of polling server tasks used to provide for execution of sporadic tasks in cyclic executive systems indicated that the period, deadline, and execution time of the polling task should be selected as $min\{(d_a - e_a - 1), mit\}$, e_a, *and* d_a, respectively. Such selection is proper in this case, as well, if the aperiodic task is a sporadic task. Since rate-monotonic priority assignment is used, the polling server task will have the highest priority. The processor utilization for the periodic task system is

$$U = \frac{2}{8} + \frac{4}{12} + \frac{2}{6} = 0.917 \leq 1.$$

The utilization bound is inconclusive, but the task system is schedulable. Figure 7.14 shows a scenario in which two aperiodic task releases occur, as indicated in the diagram, at times $t = 7$ and $t = 15$. The aperiodic task execution times are 1 and 2, respectively. The server task τ_s is shown being released at a periodic rate with period $p_a = 6$, but execution is shown only when there is an aperiodic task pending. There is some execution time required to check whether or not an aperiodic task release is pending, but this normally takes minimal time. It may actually be accomplished by the multitasking executive itself. This is a scenario, not a schedule, since it shows the execution of the task system, dynamically scheduled by a priority-driven multitasking executive responding to a specific pattern of aperiodic task releases.

The server capacity, the remaining server execution time, is shown in the diagram: it decreases linearly as the server task executes. The remaining server capacity is relinquished upon completion of any periodic task execution, and it is replenished periodically each time the server task is scheduled to execute. From $t = 12$ to $t = 13$, for example, the server capacity decreases from its initial value of 2 to a residual value of 1 as the first aperiodic release is executed. Server capacity remaining after completion of the aperiodic task is used by the multitasking executive to execute the highest priority pending task, in this case, task τ_2.

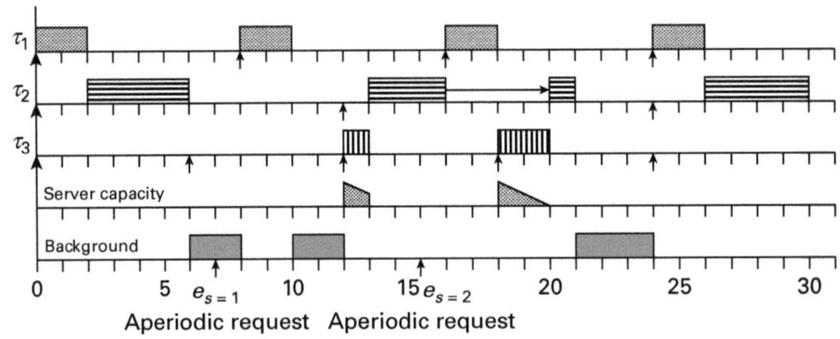

Figure 7.13 Use of a Polling Task to Execute Aperiodic Tasks.

Polling server tasks can be used to service aperiodic tasks whose releases are not announced by interrupts since the server task executes periodically as part of the periodic task system. It can, upon start-up, perform whatever checks are necessary to determine if an aperiodic task has been released.

The time from the point at which an aperiodic task is released and the time its execution begins is termed the *task latency*. For rather obvious reasons, task latency is an important issue in real time systems with hard aperiodic timing constraints. In general, the use of polling tasks to provide for the execution of aperiodic tasks may produce excessive task latency times. This effect can be seen, to some extent, in the scenario of Figure 7.13, but it is graphically illustrated in the situation depicted in Figure 7.14, in which a polling server task τ_s and periodic task τ_1 both have period of 50. The periodic task has execution time $e_1 = 49$ while the polling server task has execution time $e_s = 1$. It is clear that an aperiodic task release occurring just after the polling server has completed will have excessive latency. The average latency time in this case would be 25, assuming that the aperiodic release times are uniformly distributed.

It is to address this problem of excessive latency that can occur with use of polling tasks that the bandwidth preserving algorithms were developed. The following section introduces the basic concepts behind the development of these algorithms and presents examples of their use.

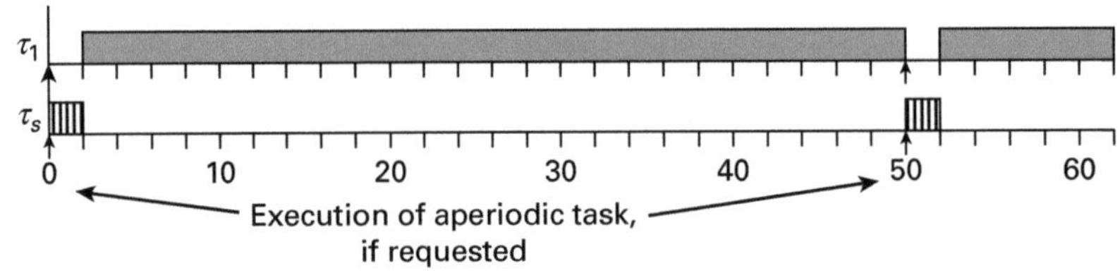

Figure 7.14 Excessive Latency Time with a Polling Server Task.

7.2.3 Bandwidth Preserving Aperiodic Task Server Algorithms

Using an interrupt handler to provide for execution of an aperiodic task produces very short latency times but can play havoc with the scheduling of the periodic task system. On the other hand, the use of a periodic task, such as a polling server, readily accommodates the scheduling of the periodic task system but can produce excessive latency times. The bandwidth preserving algorithms, so named because they preserve the aperiodic response time lost through polling, attempt to balance the requirements for small latency times and the integrity of the periodic task schedule.

7.2.4 The Deferrable Server Algorithm

The poor latency time characteristic of the use of a polling server task results from the fact that the server task relinquishes all its execution time if there is no aperiodic release pending when the server task begins execution. It also relinquishes any remaining execution time after the aperiodic task has completed. A modification to the polling server technique—called the deferrable server algorithm—preserves the execution time of the server task until the end of its period. The deferrable server algorithm functions in the following manner.

- If an aperiodic task is released, and there is available execution time allocated to the deferrable server task, the server task will preempt any lower priority task.
- This preemptive execution of the aperiodic task will continue until the end of the server task's period or until the server capacity reaches zero.
- As with the polling server, the deferrable server task's execution time is periodically replenished to its full value each time the server task is scheduled for execution.

Use of the deferrable server algorithm is not without its drawbacks. The main disadvantage is the loss of the capability for schedulability analysis, which results from the fact that the execution of the server task is no longer periodic. That is, since the server task may preempt any lower priority task, upon release of an aperiodic task, as long as there is any server capacity remaining, the server task execution is non-deterministic. The schedulability analysis of periodic task systems using a particular priority assignment presumes that when a task is the highest priority task available for execution, it must execute. The deferrable server algorithm violates this presumption and, by deferring execution of the server task, may cause a missed deadline in a lower priority task. When the deferrable server algorithm is used in conjunction with the rate-monotonic priority assignment algorithm, schedulability bounds can be developed.

A summary of the characteristics of the deferrable server algorithm is presented below.

- A deferrable server task is a high priority periodic task dedicated to servicing aperiodic tasks.
- The deferrable server algorithm improves performance over that obtained using a polling server task by preserving the execution time of the server task until it is needed for execution of an aperiodic task.
- Simulation results indicate that the use of the deferrable server algorithm improves aperiodic response times by a factor of between two and ten over that obtained with background processing.
- The performance gain over the use of a polling task depends on the period of the server task. For high frequency servers, with server rates higher than the average aperiodic task interarrival rate, the deferrable server response converges to that produced by the use of a simple polling server task. For lower frequency servers, considerable improvement can usually be expected.

- Implementation is relatively simple, requiring only some means of measuring increments of server task execution time. Many real time executives provide user definable counters that can be used for this purpose.
- Schedulability analysis is negatively affected by the fact that the server task no longer executes periodically.

7.2.5 The Priority Exchange Algorithm

While the deferrable server algorithm preserves its execution time by deferring execution until an aperiodic task is released, another bandwidth preserving algorithm, the priority exchange algorithm, preserves execution time by exchanging unneeded aperiodic execution time for lower priority periodic execution time. This algorithm functions in the following manner.

- Each time the server task is scheduled for execution, its execution time is brought to full capacity.
- At any instant of time, if aperiodic time is available, and aperiodic tasks are pending, that is, they have been released but are not currently executing, then, the aperiodic tasks are executed.
- Otherwise, the highest priority pending periodic task is executed, or background processing is begun if there is no pending periodic task.
- If a periodic task is executed at the aperiodic priority level, then a priority exchange occurs—the periodic task executes at the priority of the aperiodic server task and aperiodic server time as accumulated at the priority level of the of the task being executed.
- Any execution—even that of background processing—occurs at the highest priority level available. A task executes either at its own priority level or at the highest server capacity priority level—whichever is highest.
- Ties are always broken in favor of the aperiodic task.

The priority exchange algorithm exhibits certain advantages over the deferred server algorithm, one of which is that the ability to apply the schedulability analysis techniques is preserved.

> For any task system with static priorities and which includes a server of aperiodic tasks, the server task may exchange execution time with any lower priority task without affecting the schedulability of the periodic tasks—an exchange only advances periodic task execution at the expense of delaying aperiodic tasks. Thus, if a periodic task system is schedulable, it is also schedulable if one or more of the periodic tasks is a server task implemented using the priority exchange algorithm.

Such a statement cannot be made in the case of the deferrable server algorithm because, as noted above, a server task using the deferrable server algorithm does not execute periodically and, thus, may cause lower priority tasks to miss their deadlines.

A summary of the characteristics of the deferrable server algorithm is presented below.

- A high priority periodic task is dedicated to servicing aperiodic tasks.
- The priority exchange algorithm improves performance over that obtained through use of a polling task by exchanging server execution time for lower priority periodic execution time.

- Simulation results indicate that the priority exchange algorithm is equivalent to the deferrable server algorithm in average aperiodic task response time.
- Server utilization is higher than that obtainable by the deferred server algorithm.
- Implementation of the priority exchange algorithm is much more complex than that of the deferrable server algorithm. The accumulation of aperiodic execution time at various priority levels represents a relatively high overhead in terms of complexity and executive software execution time.
- The capability for schedulability analysis of the periodic task system is preserved.

Both the priority exchange algorithm and the deferrable server algorithm have the capability of resolving the latency time problem encountered with the use of a simple polling task. Each also has certain disadvantages—the loss of schedulability analysis in the case of the deferrable server algorithm and the implementation complexity of the priority exchange algorithm.

7.2.6 The Sporadic Server Algorithm

On the basis of the remarks above, an effective method of providing for the execution of aperiodic tasks would exhibit an aperiodic task response time equivalent to that of the deferrable server and priority exchange algorithms, would have an implementation complexity comparable to that of deferrable server algorithm, would preserve the capability of schedulability analysis of periodic task systems, and would effectively provide for execution of both hard deadline and soft deadline aperiodic tasks. The sporadic server algorithm, which is a simple modification of the deferrable server algorithm, meets these requirements.

As with the previously described bandwidth preserving algorithms, the sporadic server algorithm is characterized by a priority level at which the server task executes and an execution time available for the execution of aperiodic tasks. The sporadic server algorithm differs in the manner in which the execution time of the server task is replenished. There is no task period, in the usual sense of the term, but, rather, a replenishment period, a time duration that must pass before used execution time can be restored. A replenishment time is established dynamically by the algorithm, and at the replenishment time, the server execution time is replenished only to the extent that it has been used in the execution of aperiodic tasks. This method of replenishment has one very important effect.

> *A periodic task system that is schedulable is also schedulable if one or more of the periodic tasks is replaced by a sporadic server task with the same period and execution time.*

This is the same property, resulting in preservation of the capability for schedulability analysis, exhibited by the priority exchange algorithm. In addition, the sporadic server preserves the high priority execution time in the same manner as does the deferrable server algorithm. Furthermore, because there is no multiple priority accumulation of aperiodic execution time, as is the case with the priority exchange algorithm, the implementation complexity of the sporadic server algorithm is only slightly greater than that of the deferrable server algorithm. For these reasons, the sporadic server algorithm would seem to be the best choice for replacing the simple polling task in those situations in which aperiodic response times are critical.

There are several variations of the sporadic server algorithm, each differing in how the replenishment times are determined. The particular version of the algorithm described below is the simplest variation.

- When aperiodic execution time is available, that is, when it has not been exhausted, the sporadic server will compete for processor time at its assigned priority.
- When aperiodic execution time is exhausted, or when there are no pending aperiodic task releases, the sporadic server will relinquish the processor.
- When the sporadic server relinquishes the processor for either of the reasons above, a replenishment time is established to be one replenishment period after the sporadic server began execution.
- At the replenishment time, the amount of execution time replenished is that used by the sporadic task from the time it began execution to the time it relinquished the processor.

The implementation complexity is rather low, requiring only the use of counters to measure expended aperiodic execution time and some interaction with the real time multitasking executive.

Example 7.2.3: Use of the sporadic server algorithm.

A scenario illustrating the use of the sporadic server algorithm is depicted in Figure 7.15. The task system of this example is specified below.

$N = number\ \ of\ \ periodic\ \ tasks = 3,\ including\ \ the\ \ polling\ \ server\ \ task$

M = number of processors = 1

Period	Deadline	Execution Time
$p_1 = 8$	$d_1 = 8$	$e_1 = 2$
$p_2 = 20$	$d_2 = 20$	$e_2 = 6$
$p_s = 10$	$d_s = 10$	$e_s = 4$

The polling server task τ_s has been selected to have a replenishment period of 10 and an execution time of 4, based upon the particular characteristics of the aperiodic task(s) it is to serve. Rate-monotonic priority assignment will be used; the polling server task will execute at a medium priority. The processor utilization for the periodic task system is

$$U = \frac{2}{8} + \frac{6}{20} + \frac{4}{10} = 0.95 \leq 1.$$

The utilization bound test is inconclusive, but the task system is schedulable.

The scenario depicted in Figure 7.15 shows five aperiodic task releases, two requiring two units of execution time and three requiring four units of execution time. The server capacity is established

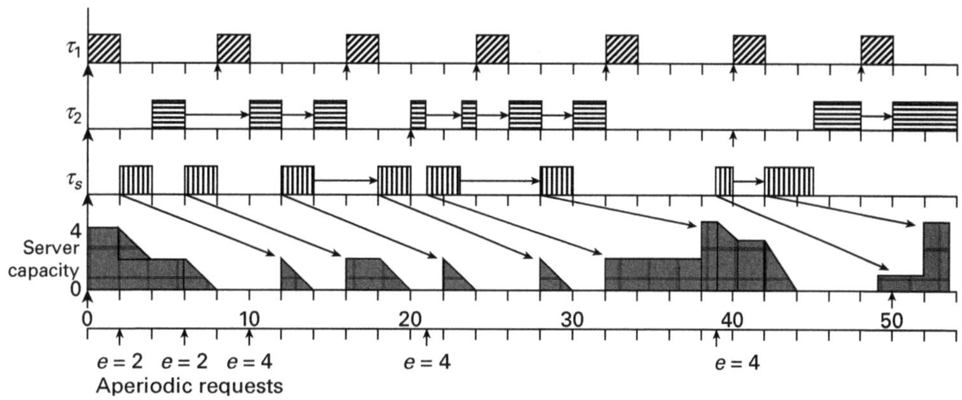

Figure 7.15 Aperiodic Task Execution Using the Sporadic Server Algorithm.

initially as the full capacity of four time units, the execution time of the server task. The server capacity time line shows the server capacity as a function of time as the task system executes. When aperiodic tasks are executing, the server capacity decreases linearly as the execution time is used. Replenishment times—indicated by *R* in the diagram, are established each time the server task relinquishes the processor.

The replenishment time is one replenishment period, in this example, ten time units, after the server task began the current execution (the *current execution* is the execution of the server task that is terminated when the server task relinquishes the processor), and the amount of server execution time replenished is that used during the current execution. It should be noted that the sporadic server task, in this example, does not execute at the highest priority, yet all aperiodic task releases are serviced with relatively small latency times.

A summary of the sporadic server algorithm is presented below.

- Server tasks can be created at any priority level without affecting the ability to perform schedulability analysis. This is not the case for the deferrable server algorithm.
- The implementation complexity of the sporadic server is relatively low. This is not the case for the priority exchange algorithm.
- The server size can be larger—contain more execution time—than a comparable server implemented using the deferrable server algorithm.
- The response time is comparable to that obtained using the deferrable server or priority exchange algorithms.

7.2.7 The Extended Priority Exchange Algorithm

Schedulability analysis presumes knowledge of the execution time e_n of each task in the task system. When the system actually executes, the execution time of a task will vary from one release to another, a characteristic known as jitter. Jitter can arise from several causes, but, basically, it represents differences in execution times caused by differing execution paths through the code comprising the task from one execution to the next. To account for this variation in execution time, schedulability analysis is often performed using worst-case execution time or, at least, a value greater than the average to offset the effects of jitter. When, in addition, the deferrable server, priority exchange, or sporadic server algorithms are used to provide for the servicing of aperiodic tasks, the resulting creation of server tasks further adds to the processor utilization requirements that must be factored into the schedulability analysis.

When worst-case execution times are used to determine the schedulability of a task system, the result is a task system execution that, on the average, has more background (idle) processing than is indicated by the schedulability analysis. The basic concept behind the extended priority exchange algorithm is to use the difference between the worst-case and average execution times of periodic tasks to service aperiodic tasks. In this manner, no server tasks are created, thus, allowing more processor utilization to be applied to providing worst-case execution times. A simple example of the concept behind the extended priority exchange algorithm can be provided by considering the behavior of the task system specified in the following example.

Example 7.2.4: Average and worst-case execution times of a simple task system.

The specification of a task system is given below. Both average and worst-case execution times are given for each of the two tasks comprising the task system.

$$N = number\ of\ tasks = 2$$

Period	Deadline	Average	Worst-case
$p_1 = 20$	$d_1 = 20$	$e_1 = 6$	$e_1 = 8$
$p_2 = 40$	$d_2 = 40$	$e_2 = 12$	$e_2 = 16$

The average and worst-case processor utilizations are:

$$U_{av} = \frac{6}{20} + \frac{12}{40} = 0.6 \qquad U_{wc} = \frac{8}{20} + \frac{16}{40} = 0.8$$

In both cases, the processor utilization is below the utilization bound for rate-monotonic scheduling, which is 1.0 for harmonically related task periods. Hence, the task system is schedulable even if worst-case execution times are realized. Task schedules resulting from the average and worst-case execution times are shown in Figure 7.16. The difference in background processing time between these two schedules is evident in the diagram of Figure 7.16, and it is this difference that the extended priority exchange algorithm uses for the processing of aperiodic tasks.

The average execution time is that predicted to be the long-term average value of task execution times as the system executes. Actual execution times for various releases of the task will vary from values smaller than the average to the worst-case value. Some major cycles will, then, have considerable excess processor capacity, appearing as background processing while others will have a minimal amount.

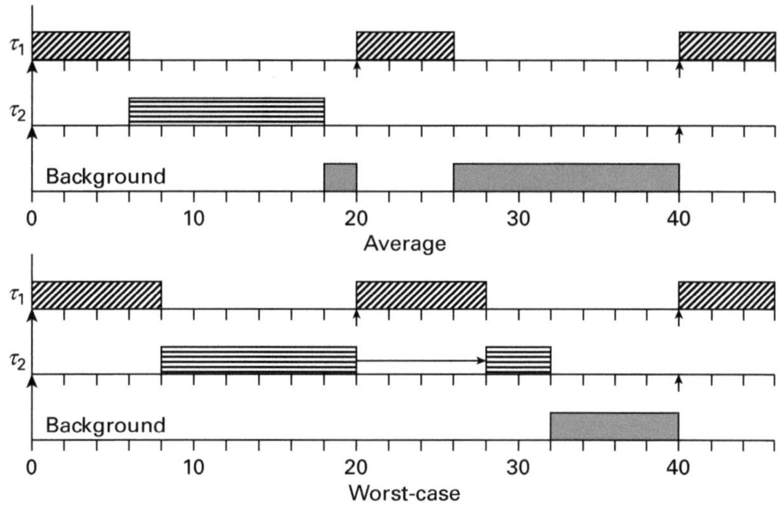

Figure 7.16 Schedules for Average and Worst-Case Execution Times.

Example 7.2.5: Average and worst-case execution times with a server task.

The specification of a task system is given below. Both average and worst-case execution times are given for each of the two tasks comprising the task system.

$N = number\ of\ tasks = 3$, *including the server task*

Period	Deadline	Average	Worst-case
$p_1 = 20$	$d_1 = 20$	$e_1 = 6$	$e_1 = 8$
$p_2 = 40$	$d_2 = 40$	$e_2 = 12$	$e_2 = 16$
$p_s = 10$	$d_s = 10$	$e_s = 2$	

The average and worst-case processor utilizations are:

$$U_{av} = \frac{6}{20} + \frac{12}{40} + \frac{2}{10} = 0.8 \qquad U_{wc} = \frac{8}{20} + \frac{16}{40} + \frac{2}{10} = 1.0$$

Both the average and worst-case execution times result in schedulable task systems since the utilization bound is 1.0 in the case of harmonically related task periods.

Task schedules for the average and worst-case execution times are illustrated in the diagram of Figure 7.17, in which the server task is shown executing for the full duration of its execution time. This situation is tantamount to a presumption that aperiodic tasks are always pending and that their execution consumes all available execution time, a limiting case used here to compare the results obtained using a server task to those obtained using the extended priority exchange algorithm.

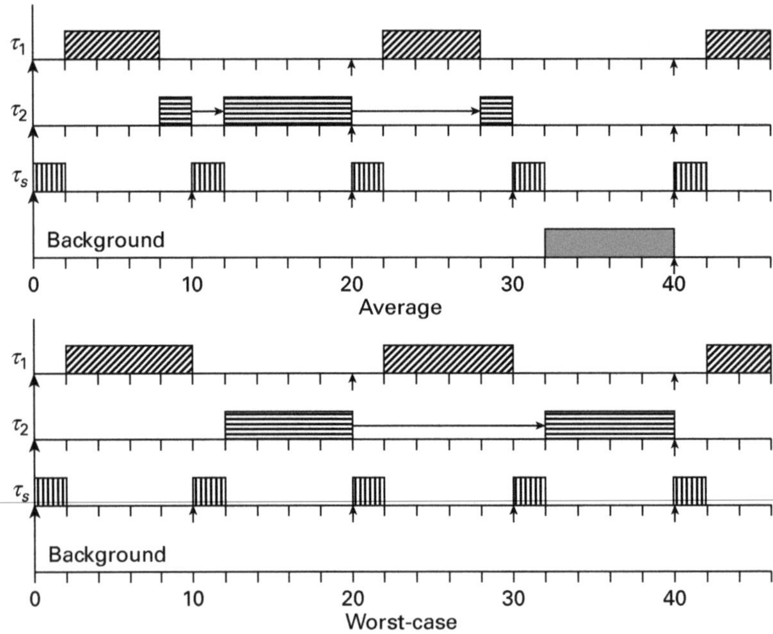

Figure 7.17 Schedules for Average and Worst-Case Execution Times with a Server Task.

As an alternative to the creation of a separate server task, application of the extended priority exchange algorithm to the task system of this example will result in the utilization, for the processing of aperiodic tasks, of the difference in background processing in the average and worst-case schedules shown in Figure 7.17.

While the concept behind the extended priority exchange algorithm is rather straightforward, the algorithm itself is fairly complex, and further discussion requires the following definitions.

- L_s is the priority level at which the system is executing at any instant of time. That is, it is the current priority of the executing task.
- L_n is the priority level of task τ_n. That is, it is the assigned priority of task τ_n.
- PTA_n is the *periodic time available*, the total time available for the execution of periodic tasks at priority level L_n.
- ATA_n is the aperiodic time available—the total time available for the execution of aperiodic tasks at priority level L_n.

The extended priority exchange algorithm requires the maintenance of values of PTA_n and ATA_n for all priority levels L_n.

The efficacy of the extended priority exchange algorithm depends upon a feature of priority-driven scheduling that ensures that as long as the maximum execution time of each priority level is not exceeded, the execution time allocated to any priority level can be *fully used* prior to the end of the associated period. The implication is that anytime a periodic task does not use its total execution time, the remaining unused execution time can be used for the processing of aperiodic tasks. The only requirement is that the aperiodic execution end at or before the end of the periodic task's period. Since the task system specification is based on worst-case execution times rather than average execution times, the execution of any given task for *most releases* will result in this kind of unused execution time. The priority exchange algorithm provides for the allocation of this unused execution time in a manner that preserves the schedulability of the periodic task system.

Finally, if the extended priority exchange algorithm is to have access to the maximum amount of excess execution time, the task system worst-case execution times should be artificially increased to the maximum possible value for which schedulability is maintained. For example, the worst-case task schedule of Example 7.2.4, as shown in Figure 7.16, still has eight units of unused execution time appearing as background processing. Those eight units could be assigned to task τ_1 by increasing its worst-case execution time from eight units to twelve units, thus, increasing the worst-case processor utilization to 1.0 and still be within the schedulability bound for this task system. In this manner, additional high priority periodic execution time is made available for use by the priority exchange algorithm. This procedure is generalized below.

- Using the established priority levels of the task system, the worst-case periodic execution times are estimated, and from these, the utilization of each priority level is determined.
- The utilization of each priority level—beginning with the highest priority level, is artificially increased by increasing the worst-case execution times of the task or tasks with that assigned priority until the limit of schedulability of the task system is reached. Starting with the highest priority level ensures that the additional utilization occurs at the highest possible priority. The limit of schedulability is determined using the schedulability analysis techniques discussed earlier.

This process can be quite complex for general priority lists since determining the limit of schedulability might entail an iterative process of schedule construction or evaluation of utilization bounds. When rate-monotonic priority assignment is used, the process is much simpler since the easily applied schedulability bounds can be used.

The procedure for implementing the extended priority exchange algorithm is described below. It is presumed that the task system with possibly artificially increased worst-case execution times has been shown to be schedulable using the prescribed priority assignment.

- Whenever a periodic task is released, the periodic time available (*PTA*) for the corresponding priority level is increased by an amount equal to the worst-case execution time of the task.
- At any instant of time, the currently executing task—periodic or aperiodic—executes at the highest priority level that has either non-zero *PTA* or non-zero aperiodic time available (*ATA*).
- Periodic tasks qualify for execution using their assigned rate-monotonic priorities and can use either *PTA* or *ATA* of a higher priority level if it is available. Aperiodic tasks use only *ATA* .
- Whenever periodic execution time at a given priority level is used for task execution, the *PTA* for that priority level is reduced accordingly.
- Whenever aperiodic execution time at a given priority level is used for task execution, either periodic or aperiodic, the *ATA* for that priority level is reduced accordingly.
- When an aperiodic task is released and the highest priority level with execution time available, either *PTA* or *ATA*, has non-zero *ATA*, the aperiodic task is executed at that priority level. Otherwise, the highest priority pending periodic task is executed at that priority level.

- Whenever *ATA* at a given priority level is reduced to zero, execution of the aperiodic task at that priority level is suspended.
- When a periodic task executes at its assigned priority level, the *PTA* value for that level is reduced accordingly.
- If a periodic task executes at a priority level higher than its assigned priority, then the *ATA* value of the higher priority level is reduced accordingly, and at the assigned priority level of the task, the value of *PTA* is reduced and the value of *ATA* increased by a like amount.
- When a periodic task completes, the execution time used to complete task, which will vary from one release of the task to another, is compared to the worst-case execution time used in the schedulability analysis.

  If the difference is non-zero, the value of the *ATA* at the priority level of the task is increased by the amount of the difference, and the value of the *PTA* at that priority level is decreased by a like amount.
- Aperiodic always wins priority ties when *ATA* at any level is non-zero

In essence, the extended priority exchange measures the difference between the worst-case execution times used in the schedulability analysis and the actual execution times for each release of the periodic tasks and, then, allocates that time difference to the execution of aperiodic tasks. The complexity of algorithm results from the requirement for executing both periodic and aperiodic tasks in a manner that does not affect the schedulability of the periodic task system.

Example 7.2.6: Use of the Extended Priority Exchange algorithm.

If the extended priority exchange algorithm is applied in the case of the task set of Example 7.2.5 under the conditions presumed in that example, that aperiodic tasks are always pending and consume all available execution time, the resulting task schedule is depicted in Figure 7.18. The first step in the application of the extended priority exchange algorithm is to ensure that the periodic task system is schedulable using the worst-case execution times. This determination was made in Example 7.2.4 by showing that the worst-case processor utilization is less than the

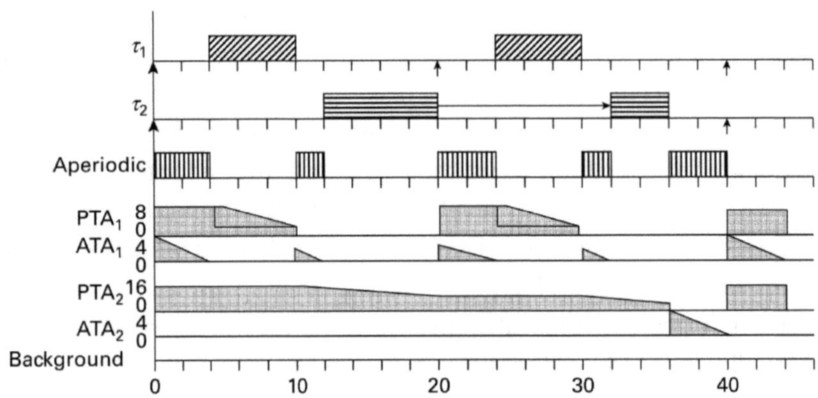

Figure 7.18 Scheduling Aperiodic Task Execution Using the Extended Priority Exchange Algorithm.

rate-monotonic bound, which is 1.0, in this case. The time line labeled *aperiodic* in the diagram of Figure 7.18 represents actual execution of aperiodic tasks. Those labeled PTA_1, ATA_1, PTA_2, and ATA_2 are the values of these categories of execution time as they must be maintained by the run time support for the extended priority exchange algorithm.

A summary of the extended priority exchange algorithm is given below.

- No dedicated periodic server task is created; instead, unused periodic execution time is used for execution of aperiodic tasks.
- The extended priority exchange algorithm is effective even when the utilization of the periodic tasks is so high that there is none available for creating a schedulable server task.
- Monitoring of processor time utilization in a variety of categories is required, making the implementation overhead high.
- Simulation results indicate that use of the extended priority exchange algorithm can provide better aperiodic response than those techniques using server tasks, even in situations in which the worst-case and average utilizations differ by only a relatively small amount.
- There is no guarantee that sufficient processor time will always be available when it is required by an aperiodic task.

In dynamically scheduled systems, asynchronous processing in the form of aperiodic tasks is often incorporated into the periodic schedule through the use of server tasks—periodic tasks placed into the periodic schedule solely for the purpose of providing execution time to aperiodic tasks.

7.2.8 Summary of Aperiodic Task Server Algorithms

Background processing may be used to provide execution time for aperiodic tasks if the processing is not required to meet a hard deadline. A polling task may be used to provide better response than background processing. The deferrable server algorithm, the priority exchange algorithm, and the sporadic server algorithm all call for the creation of periodic server tasks and offer better response time than simple polling.

- The sporadic server, in particular, provides good response time with relatively simple implementation.
- The extended priority exchange algorithm uses a rather novel technique to provide execution time to aperiodic tasks. It uses the difference between the worst-case execution times, used in schedulability analysis, and the average execution times actually experienced as the task system executes, for the execution of aperiodic tasks. The algorithm provides good response time for aperiodic tasks, but its implementation is complex.
- A practical combination of real time task-scheduling techniques is the use of a static priority assignment: deadline-monotonic or rate-monotonic together with the priority ceiling protocol as the synchronization protocol and the sporadic server algorithm as the method of providing execution time to aperiodic tasks.

It is important to the real time systems development process that software systems engineers have an appreciation for the general complexity of the task scheduling and synchronization problems. Tuning problems can easily be the primary cause of software failure in real time systems. In general, analytic techniques can play a vital role in the design process by providing the real time systems engineer with an understanding of the timing characteristics of the system under design, with a method to analyze and predict system timing characteristics and to evaluate the effect of alternative design approaches on the timing behavior of the system.

APPENDIX A

NUMBERING SYSTEMS

Humans communicate with letters and numbers. Computers communicate in electrical signals—very low voltage electrical signals that have only two possible states. These electrical states may be different in any particular application. Particular levels may be −0.5 volts and 1 volt, 0 volts and +1.0 volts, +3 volts and +5 volts, or any others as long as they can be differentiated from one another. The actual values are determined by the technology used in the production of the electrical elements themselves. But it doesn't make much difference (except for the switching speed of the elements) what the voltages are because the computer recognizes only two values, high and low, zero and one, on or off, etc. These values are called *bits*, and they can only have two values, zero or one.

Some time ago, humans developed a number system that had only two values, 0 and 1. This was a perfect fit to the manner in which computers deal with data. For example, the sequence of bits 11101011 to a computer represents the decimal number 235. And, conversely, the decimal number 235 to a computer represents the binary number 11101011. The choice of eight bits in this example is significant because the grouping of eight bits is called a *byte*. A grouping of bytes is called a *word*. At this point in the development of the modern computer, operators had to convert operations, such as *ADD*, into the binary format, or *instruction*, that the computer would recognize as a request to perform an add operation. The data also had to be converted to a binary form for the computer to perform the required operation.

The scenery in the computer room saw an operator manually translating expressions like *ADD* into a binary sequence of bits and feeding that to the computer. To make things easier for the operator, instructions and data were expressed in bytes. Very quickly, computer instructions were expressed as human words, like *ADD*, with the responsibility for translating placed on the computer itself. Then, the operator, or user, could write programs in something like human language. The computer program that provided this service was called an assembler.

The next stage was the development of programs that could accept a program written in very human-like language and which relieved the programmer of all the tedious work of translating it into the series of bytes that the computer could understand. The first of these so-called high level languages developed for scientific purposes was FORTRAN. The C and C++ programming languages have evolved as the preferred real time languages. Since the introduction of FORTRAN, a myriad of special-purpose languages has been developed, but the underlying operations still rely on bits, bytes, and binary words.

A.1 Number Systems

The mathematics described above are part of a much larger system of treating numbers. The *binary* system is called a *base* –2 system. Other number system commonly found in the computer world are *base* –8 (octal) and *base* –16 (hexadecimal). The manner in which these various number systems are handled is identical. Good thing, since theoretically there are an infinite number of numbering systems.

Base-2	Binary
Base-3	Ternary
Base-4	Quaternary
Base-5	Quinary
Base-6	Senary
Base-7	Septenary
Base-8	Octal
Base-9	Nonary
Base-10	Decimal
Base-11	Undenary
Base-12	Duodecimal
....	
Base 16	Hexadecimal
....	

and so on.

Table A.1 Comparison of Values in Different Bases.

Decimal Value	Binary Value	Octal Value	Hexadecimal Value
0	000000	00	00
1	000001	01	01
2	000010	02	02
3	000011	03	03
4	000100	04	04
5	000101	05	05
6	000110	06	06
7	000111	07	07
8	001000	10	08
9	001001	11	09
10	001010	12	0A
11	001011	13	0B
12	001100	14	0C
13	001101	15	0D
14	001110	16	0E
15	001111	17	0F

No matter what decimal number is selected, there is a binary number representing the same value. So is an octal number, a hexadecimal number, a nonary number, and so on. Moreover, the conversion of a binary number in one base to the same number in a different base is fairly straightforward.

Humans feel comfortable with *base* − 10 (decimal) numbers (probably leading back to ten fingers). Computers, on the other hand, prefer octal and hexadecimal because of computational advantage gained through the use of harmonically related numbers (2, 4, 8, 16).

Computer engineers often need to write out binary quantities, but, in practice, writing out a binary number such as 1001001101010001 is tedious and prone to errors. Therefore, binary quantities are written in a *base* − 8, or *octal*, or much more commonly, a *base* − 16, *hexadecimal* or *hex,* number format. In the *decimal* system, there are 10 digits, 0 through 9, which combine to form numbers. That is, the value of an octal "10" is the same as a decimal "8," an octal "20" is a decimal "16," and so on. In the octal system, there are eight digits (bits in the current context). In the decimal system, there are ten digits, and in the hexadecimal system, there are sixteen digits. In the case of hexadecimal and its sixteen digits, six additional digits, A, B, C, D, E, and F are added, as shown in Table A.1.

It is important to realize that reading across this table—for example, 11, 001011, 13, and 0B are all the same number.

A.2 Conversion Between Bases

The discussion here deals only with systems using positional notation. That is, the value of the number depends on the position of the digits. For example, the number 273 has a different value than the number 732 even though both numbers contain the same digits. In a positional number system, the value of each digit is determined by which place it appears in the full number. The lowest place value is the right-most position, and each successive position to the left has a higher place value.

In the decimal system, the right-most digit is the "*one*" position. The digit to its left is the "*tens*" position. To the left of the "*tens*" digit is the "*hundreds*," and so on. Then, the decimal number 732 can be written as $7 \times 100 + 3 \times 10 + 2 \times 1$. More generally, it can be expressed as

$$(7 \times 10^2) + (3 \times 10^1) + (2 \times 10^0) = 732_{10} \qquad (10^0 = 1)$$

A similar relationship exists for any other number system. In octal, the number 732_8 can be expressed as

$$(7 \times 8^2) + (3 \times 8^1) + (2 \times 8^0) = 474_{10} \qquad (8^0 = 1)$$

The expression above actually is a conversion of the octal value 732_8 to its decimal value 474_{10}.

(The subscript notation 732_8 means the octal number 732. Similarly, 474_{10} represents the decimal number 474. It is customary to discard the subscript if it is ten.)

The process above holds true for the conversion of a decimal number to any base number system. The process is to repeatedly divide the decimal number by the base (radix) to which it is to be converted until the quotient becomes zero. As the number is divided, the remainders, in reverse order, form the digits of the number in the new base. For example, to convert the decimal number 82 to its senary (*base*−6) representation:

$82/6 = 13$ remainder 4

$13/6 = 2$ remainder 1

$2/6 = 0$ remainder 2

Taking the remainders in reverse order: Answer is 214_6

A.3 Negative Values of Binary Numbers

From this point on, only binary numbers will be considered since the primary focus relates to the use of numbers in computers. The coverage, to this point, deals only with positive binary numbers.

In computers, the usual method for storing signed numbers uses a form called the *two's complement*, which is formed by changing all zeros to ones and all ones to zeros and, then, adding one. For example, to form the two's complement of the decimal number 53:

53 −> 00110101

two's complement −> 11001010 + 1 = 11001011

Some sample values of two's complement are shown in Table A.2 below.

Table A.2 Two's Complement Examples.

Bits	Unsigned value	Two's complement value
1111 1111	255	−1
1111 1110	254	−2
1000 0010	130	−126
1000 0001	129	−127
1000 0000	128	−128
0111 1111	127	127
0111 1110	126	126
0000 0000	0	0
0000 0010	2	2
0000 0001	1	1

In two's complement representation, positive numbers are simply represented as themselves, and negative numbers are represented by the two's complement of their absolute value. The sum of a number and its two's complement will always equal zero (the last digit is truncated). The two's complement system has the advantage in that the fundamental mathematical operations of addition, subtraction, and multiplication are identical to those for unsigned binary numbers as long as the inputs are represented by the same number of bits, and any overflow beyond that number of bits is discarded from the result.

This property makes the system both simpler to implement and capable of easily handling higher precision arithmetic.

A.4 Convert Fractional Decimal Numbers to Fractional Binary Numbers

So far, the discussion has addressed only whole numbers. What about fractional binary numbers? There are methods for converting a given fractional decimal number to its equivalent binary fraction. The process described here is termed the *successive multiplication method*. The process is as follows:

1. The fractional part of the given decimal number is multiplied by 2 (i.e., the radix of binary number system).
2. The product obtained has an integer part and fractional part; the integer part here is also referred to as a carry.
3. The carry that is produced at each multiplication iteration becomes a digit in the fractional binary number.
4. The fractional part obtained is again multiplied, and the process is repeated until the fractional part becomes zero, or the number of multiplication iterations equals the number of significant digits after the decimal point in the given fractional decimal number.
5. The carry produced at each stage is taken from the first iteration to the last iteration to form the numerals in the fractional binary number, i.e., the carry obtained in the first multiplication iteration is the most significant bit (MSB) after the decimal point, and the carry obtained in the last multiplication iteration is the least significant bit (LSB).

This procedure is illustrated in the following example.

Convert $(0.625)_{10}$ decimal number to binary number using the successive multiplication method.

1st Multiplication Iteration
Multiply 0.625 by 2

$0.625 \times 2 = 1.25$ (Product) Fractional part $= 0.25$ Carry = **1** **(MSB)**

<u>2nd Multiplication Iteration</u>
Multiply 0.25 by 2

$0.25 \times 2 = 0.50$ (Product) Fractional part $= 0.50$ Carry $= 0$

<u>3rd Multiplication Iteration</u>
Multiply 0.50 by 2

$0.50 \times 2 = 1.00$ (Product) Fractional part $= 1.00$ Carry $=$ **1** **(LSB)**

The fractional part in the 3rd iteration becomes zero and, hence, the iterations are terminated. The carry from the 1st multiplication iteration becomes MSB, and the carry from 3rd iteration becomes LSB. Hence, the fractional binary number of the given fractional decimal number $(0.625)_{10}$ is $(0.101)_2$.

A.4.1 Convert Fractional Binary Numbers to Fractional Decimal Numbers

The process for conversion of fractional binary numbers to fractional decimal numbers should be evident at this point. Two examples of this conversion are shown below.

Problem 1: Convert $(101.101)_2 = (\ ?\ )_{10}$
$= 1\,0\,1\,.\,1\,0\,1$
↑ ↑
MSB LSB

$$= 1 \times 2^2 + 0 \times 2^1 + 1 \times 2^0 + 1 \times 2^{-1} + 0 \times 2^{-2} + 1 \times 2^{-3}$$
$$= 1 \times 4 + 0 \times 2 + 1 \times 1 + 1 \times (1/2) + 0 \times (1/4) + 1 \times (1/8)$$

$$= 4 + 0 + 1 + (1/2) + 0 + (1/8)$$

$$= 5 + 0.5 + 0.125$$

$$= 5.625$$

Therefore
$(1\,0\,1\,.\,1\,0\,1)_2 = (5.625)_{10}$

Problem 2: Convert $(0.0001)_2 = (\ ?\ )_{10}$
$= 0\,.\,0\,0\,0\,1$
↑ ↑
MSB LSB

$$= 0 \times 2^0 + 0 \times 2^{-1} + 0 \times 2^{-2} + 0 \times 2^{-3} + 1 \times 2^{-4}$$
$$= 0 \times 1 + 0 \times (1/2) + 0 \times (1/4) + 0 \times (1/8) + 1 \times (1/16)$$

$= 0 + 0 + 0 + 0 + (1/16)$

$= 0 + 0.0625$

$= 0.0625$

Therefore

$(0.0001)_2 = (0.0625)_{10}$

A.5 Why Use Binary?

At the lowest level, computers are based on billions of electrical elements that have only two states (usually low and high voltage). By interpreting these as 0 or 1, it's very easy to build circuits for storing binary numbers and doing calculations with them.

While it's possible to simulate the behavior of decimal numbers with binary circuits, as has been implied by the previous discussion of binary numbers, it is less efficient than the use of binary numbers. If computers used decimal numbers internally, they would have less memory and be slower at the same level of technology.

Since the difference in behavior between binary and decimal numbers is not important for most applications, the logical choice is to build computers based on binary numbers and live with the fact that some extra care and effort are necessary for applications that require decimal-like behavior.

For real time software developers, however, there are many applications that require the use of binary numbers for the increased efficiency they provide. This kind of requirement may arise for reasons of execution time and/or cost when designing software for a product that will be required in large numbers.

A.6 The Hexadecimal Numbering System

In mathematics and computing, hexadecimal (also *base*−16, or *hex*) is a positional numeral system with a radix, or base, of 16. It uses sixteen distinct symbols, most often the symbols 0−9 to represent values zero to nine and A, B, C, D, E, F (or, alternatively, a, b, c, d, e, f) to represent values ten to fifteen. Hexadecimal numerals are widely used by computer system designers and programmers. Several different notations are used to represent hexadecimal constants in computing languages; the prefix "0x" is widespread due to its use in UNIX and C (and related operating systems and languages). Alternatively, some authors denote hexadecimal values using a suffix or subscript. For example, one could write 0x2AF3 or $2AF3_{16}$, depending on the choice of notation.

As an example, the hexadecimal number $2AF3_{16}$ can be converted to an equivalent decimal representation. Observe that $2AF3_{16}$ is equal to a sum of (2000_{16} + $A00_{16}$ + $F0_{16}$ + 3_{16}) by decomposing the numeral into a series of place value terms. Converting each term to decimal, one can further write:

$$2AF3_{16} = (2_{16} \times 16^3) + (A_{16} \times 16^2) + (F_{16} \times 16^1) + (3_{16} \times 16^0)$$
$$= (2 \times 4096) + (10 \times 256) + (15 \times 16) + (3 \times 1)$$
$$= (10{,}995)$$

Each hexadecimal digit represents four binary digits (bits). The primary use of hexadecimal notation is a human-friendly representation of binary-coded values. One hexadecimal digit represents a nibble (4 bits), which is half of an octet or byte (8 bits). For example, byte values can range from 0 to 255 (decimal) but may be more conveniently represented as two hexadecimal digits in the range 00 to FF. Hexadecimal is also commonly used to represent computer memory addresses.

A.7 Floating Point Numbers

Many embedded applications can be developed using binary numbers. This is particularly true today when most embedded systems are built around 32 bit CPUs. These devices provide a rich capability in performing the arithmetic processing required of many of today's embedded systems. Calculations can be performed on signed or unsigned integers, and 32 bits gives a good range of representable values. Binary representation is a simple yet very powerful way to represent fractional numbers in a computer. By reusing all integer arithmetic circuits of a computer, binary arithmetic is orders of magnitude faster than floating point arithmetic. This is the reason it is being used in many game and DSP applications. On the other hand, it lacks the range and precision that floating point number representation offers.

While binary numbers offer an effective path for data interchange between humans and computers, there is a downside to their use. Though they exactly represent certain numbers (those exactly expressible by the number of bits representing the number), their representations are in error for other numbers—a factor termed *overflow or underflow*. For example, a 16-bit binary number can only represent numbers in the range $-32{,}768$ to $32{,}767$. A 32-bit binary number can only represent the range $-2{,}147{,}483{,}648$ to $2{,}147{,}483{,}647$. Any decimal number outside these ranges cannot be represented in 16-bit or 32–bit binary systems.

There are other drawbacks, as well. For example, there is a trade-off between resolution and range. For a 16-bit binary number, if the resolution is to be decimal $1/2$ (instead of one), then the range is reduced to $-13{,}384$ to $13{,}383$.

And of course, there is the problem of converting *base*-10 (human speak) to *base*-2 binary words. This task would probably be assigned to the computer itself. In other words, binary numbers, while easy on the computer, can be a task for the programmer.

A.7.1 Why Floating Point?

Although there are many embedded applications that can live with the limitations imposed by binary arithmetic, there are others that require a much wider range of capabilities. Based on the discussion above, what is needed is a system using some fixed number of bits but which can represent a very large range of decimal digits and with a resolution suitable for scientific use.

The solution has been the use of *floating point* numbers. In computing, floating point is a representation that approximates a real number in a manner that supports a trade-off between *range* and *precision*. A number is represented approximately to a fixed number of significant digits (the *significand*) and is scaled using an exponent in some fixed base. The standard today for the fixed base is ten. The term *floating point* refers to the fact that a number's *radix point* can float. That is, it can be placed anywhere relative to the significant digits of the number. This position defines the exponent component of the number.

A floating point system can be used to represent, with a fixed number of digits, numbers of different orders of magnitude. A perfect example is the physicist who needs to use the speed of light (299,792,458 meters per second) and Newton's gravitational constant (0.0000000000667408) in the same equation.

To satisfy the physicist, a number format must provide accuracy for numbers at very different magnitudes. However, only *relative* accuracy is needed. To satisfy the physicist, it must be possible to do calculations that involve numbers with vastly different magnitudes.

A.7.2 The Mechanics of Base Two Floating Point

The basic idea of floating point arithmetic is to compose a number of two main parts:

- A *significand* that contains the number's digits (bits). A negative significand represents a negative number.
- An *exponent* that specifies where the decimal point is placed relative to the beginning of the significand. Negative exponents represent numbers that are very small.

The combination above can represent numbers at vastly different magnitudes, limited by the length of the exponent. Furthermore, the length of the significand specifies the resolution between two adjacent numbers.

Decimal floating point numbers are usually presented in the form of scientific notation with an explicit point between the first and second digits. For example, the speed of light, as presented to the human user, would be something like 2.99792×10^5.

A.7.3 The IEEE 754 Standard

There are a number of ways to implement floating point numbers through varying values of the significand and exponent. While various schemes have been used in the past, today, nearly all hardware and programming languages use floating point numbers in the same binary formats. These formats are described in the *IEEE* 754 Standard. The usual formats are 32 bits or 64 bits in total length.

Table A.3

	Single Precision	Double Precision
TOTAL BITS	32	64
SIGNIFICAND BITS	23 + 1 sign bit	52 + 1 sign bit
EXPONENT BITS	8	11
SMALLEST NUMBER	1.2×10^{-38}	5.0×10^{-324}
LARGEST NUMBER	3.4×10^{38}	1.8×10^{308}

- The *actual bit sequence* is the sign bit first, followed by the exponent, and, finally, the significand bits. (The significand is also called the mantissa.)
- The exponent does not have a sign; instead, an *exponent bias* is subtracted from it (127 for single and 1023 for double precision). This, and the bit sequence, allow floating point numbers to be compared and sorted correctly even when interpreting them as integers.
- The significand's most significant digit is omitted and assumed to be 1, except for *subnormal numbers* which are marked by an all-0 exponent and allow a number range beyond the smallest and largest numbers given in the table above at the cost of precision.
- There are separate *positive and a negative zero* values, differing in the sign bit, where all other bits are 0. These must be considered equal even though their bit patterns are different.
- There are special *positive and negative infinity* values where the exponent is all 1-bits, and the significand is all 0-bits. These are the results of calculations where the positive range of the exponent is exceeded or division of a regular number by zero.
- There are special *not a number* (or NaN) values where the exponent is all 1-bits, and the significand is *not* all 0-bits. These represent the result of various undefined calculations (like multiplying 0 and infinity, any calculation involving a NaN value, or application-specific cases). Even bit-identical NaN values must *not* be considered equal.

A.7.4 Conversion From Floating Point to Binary (32 bit)

The floating point number in $base-10$ has to be first converted to a binary number using the procedure from section A4. The fractional binary number is, then, represented as a normalized binary number in the form below.

$$sign * mantissa * 2^{exponent}$$

$$+\ 1.000010001 * 2^{+5}$$

The sign bit is stored in bit 32 and is 0 for positive numbers and 1 for negative numbers. The exponent can be computed by subtracting 127 which is known as the $excess-127$ form to represent negative numbers. In the normalized form of a binary number, the digit before the binary point will ALWAYS be a 1. Based on this observation, only the fractional part of the mantissa is stored.

Problem 3: Convert $+(2.75)_{10}$ to IEEE 754 32-bit floating point representation.

$$+(2.75)_{10} = +(1\,0\,.\,1\,1)_2$$

Normalized binary number representation is

$$+(1\,.\,0\,1\,1)_2 * 2^{+1}$$

Sign bit $= 0$

Exponent $= 1_{10} + 127_{10} = 128_{10} = 10000000_2$

Mantissa = 011 0000 0000 0000 0000 0000

The floating point representation is

0 10000000 01100000000000000000000

Most CPUs have special floating point registers to store numbers in this format as well as a floating point unit (FPU) that can read numbers from the floating point registers and perform mathematical operations on them and store the result in either single or double precision format.

A.7.5 Floating Point Complications

Every decimal integer can be exactly represented by a binary integer; however, this is not true for fractional numbers. In fact, every number that is irrational in $base-10$ will also be irrational in any system with a base smaller than 10. For binary, in particular, only fractional numbers that can be represented in the form p/q, where q is an integer power of 2, can be expressed exactly with a finite number of bits. Even common decimal fractions, such as decimal 0.0001, cannot be represented exactly in binary (0.0001 is a repeating binary fraction with a period of 104 bits!). This is the reason the following will not print the value 1.0000 (correct value) but, rather, something like 1.000054.

```
SUM = 0
FOR I% = 1 TO 10000
SUM = SUM + 0.0001
NEXT I%
PRINT SUM
```

The error is due to the small error in representing 0.0001 in binary propagating to the SUM.

It is very important to realize that any binary floating point system can represent only a finite number of floating point values in *exact form*. All other values must be approximated by the closest representable value. The IEEE standard specifies the method for rounding values to the "closest" representable value.

BIBLIOGRAPHY

Aho, Alfred, Ullman, Jeffrey, and Hopcropt, John. *Data Structures and Algorithms*. Pearson. 1983.

Andersson, Björn, Sanjoy Baruah, and Jan Jonsson. "Static-priority scheduling on multiprocessors." *Real-Time Systems Symposium*, 2001. (RTSS 2001). Proceedings. 22nd IEEE. IEEE, 2001.

Audsley, Neil C., et al. "Real-time scheduling: the deadline-monotonic approach." in Proc. IEEE Workshop on Real-Time Operating Systems and Software. 1991.

Audsley, Neil C. *Optimal priority assignment and feasibility of static priority tasks with arbitrary start times*. University of York, Department of Computer Science, 1991.

Audsley, Neil, et al. "Applying new scheduling theory to static priority preemptive scheduling." *Software Engineering Journal* 8.5 (1993): 284–292.

Audsley, Neil C., et al. "Fixed priority preemptive scheduling: An historical perspective." *Real-Time Systems* 8.2–3 (1995): 173–198.

Barr, Michael. *Programming Embedded Systems in C and C++.* O'Reilly Media. 1999.

Baruah, Sanjoy K., Aloysius K. Mok, and Louis E. Rosier. "Preemptively scheduling hard-real-time sporadic tasks on one processor." *Real-Time Systems Symposium*, 1990. Proceedings., 11th. IEEE, 1990.

Baruah, Sanjoy K., Louis E. Rosier, and Rodney R. Howell. "Algorithms and complexity concerning the preemptive scheduling of periodic, real-time tasks on one processor." *Real-Time Systems* 2.4 (1990): 301–324.

Bini, Enrico, and Giorgio C. Buttazzo. "Measuring the performance of schedulability tests." *Real-Time Systems* 30.1–2 (2005): 129–154.

Burchard, Almut, et al. "New strategies for assigning real-time tasks to multiprocessor systems." *IEEE Transactions on Computers* 44.12 (1995): 1429–1442.

Buttazzo, Giorgio. *Hard real-time computing systems: predictable scheduling algorithms and applications.* Vol. 24. Springer Science & Business Media, 2011.

Baker, Theodore P., and Alan Shaw. "The cyclic executive model and Ada." *Real-Time Systems* 1.1 (1989): 7–25.

Baker, Theodore P. "Stack-based scheduling of realtime processes." *Real-Time Systems* 3.1 (1991): 67–99.

Burns, Alan, and Andrew J. Wellings. *Real-time systems and programming languages: Ada 95, real-time Java, and real-time POSIX.* Pearson Education, 2001.

Carpenter, John, et al. "A categorization of real-time multiprocessor scheduling problems and algorithms." *Handbook on Scheduling Algorithms, Methods, and Models* (2004): 30–1.

Cheng, Albert MK. *Real-time systems: scheduling, analysis, and verification*. John Wiley & Sons, 2003.

Chetto, Houssine, and Maryline Chetto. "Some results of the earliest deadline scheduling algorithm." *IEEE Transactions on Software Engineering* 15.10 (1989): 1261.

Coffman Jr, Ass Prof EG, and Ronald L. Graham. "Optimal scheduling for two-processor systems." *Acta Informatica* 1.3 (1972): 200–213.

Davis, Robert I., and Alan Burns. "A survey of hard real-time scheduling for multiprocessor systems." *ACM Computing Surveys* (CSUR) 43.4 (2011): 35.

Dertouzos, Michael L., and Aloysius K. Mok. "Multiprocessor online scheduling of hard-real-time tasks." *IEEE Transactions on Software Engineering* 15.12 (1989): 1497–1506.

Dhall, Sudarshan K., and Chung Laung Liu. "On a real-time scheduling problem." *Operations Research* 26.1 (1978): 127–140.

Edwards, Stephen, et al. "Design of embedded systems: Formal models, validation, and synthesis." Readings in Hardware/Software Co-Design 86 (2001).

Fan, Xiaocong. *Real-Time Embedded Systems: Design Principles and Engineering Practices*. Waltham, MA. Elsevier Newnes. 2015.

Gajski, Daniel D., et al. *Specification and design of embedded systems*. Englewood Cliffs: Prentice Hall, 1994.

Graham, Ronald L. "The combinatorial mathematics of scheduling." *Scientific American* 238.3 (1978): 124–132.

Henzinger, Thomas A., and Joseph Sifakis. "The discipline of embedded systems design." *Computer* 40.10 (2007): 32–40.

Jeffay, Kevin, Donald F. Stanat, and Charles U. Martel. "On non-preemptive scheduling of period and sporadic tasks." *Real-Time Systems Symposium*, 1991. Proceedings., Twelfth. IEEE, 1991.

Joseph, Mathai, and Paritosh Pandya. "Finding response times in a real-time system." *The Computer Journal* 29.5 (1986): 390–395.

Kernighan, Brian W., and Dennis M. Ritchie. *The C programming language*. Upper Saddle River, NJ. Prentice Hall, 1988.

Khemka, Ashok, and R. K. Shyamasundar. "Multiprocessor scheduling of periodic tasks in a hard real-time environment." *International Journal of High Speed Computing* 5.04 (1993): 617–636.

Klein, Mark, et al. *A Practitioner's Handbook for Real-Time Analysis: Guide to Rate Monotonic Analysis for Real-Time Systems.* Springer Science & Business Media, 2012.

Kopetz, Hermann. *Real-time systems: design principles for distributed embedded applications*. Springer Science & Business Media, 2011.

Laplante, Phillip A., and Seppo J. Ovaska. *Real-time systems design and analysis: tools for the practitioner*. John Wiley and Sons, 2011.

Layland, James W. "Scheduling Algorithms for Multiprogramming in a Hard-Real-Time Environment." Readings in Hardware/Software Co-Design (2001): 179.

Lee, Edward Ashford, and Sanjit Arunkumar Seshia. *Introduction to embedded systems: A cyber-physical systems approach*. Lee & Seshia, 2011.

Lehoczky, John P. "Enhanced aperiodic responsiveness in hard-real-time environments." Proceedings of the IEEE Symposium on Real-Time Systems. 1987.

Lehoczky, John, Lui Sha, and Ye Ding. "The rate monotonic scheduling algorithm: Exact characterization and average case behavior." *Real-Time Systems Symposium*, 1989., Proceedings. IEEE, 1989.

Lehoczky, John P. "Fixed Priority Scheduling of Periodic Task Sets with Arbitrary Deadlines." *RTSS*. Vol. 90. 1990.

Lehoczky, John P., and Sandra Ramos-Thuel. "An optimal algorithm for scheduling soft-aperiodic tasks in fixed-priority preemptive systems." *Real-Time Systems Symposium*, 1992. IEEE, 1992.

Leung, Joseph Y-T. "A new algorithm for scheduling periodic, real-time tasks." *Algorithmica* 4.1–4 (1989): 209–219.

Leung, Joseph Y-T., and M. L. Merrill. "A note on preemptive scheduling of periodic, real-time tasks." *Information Processing Letters* 11.3 (1980): 115–118.

Leung, Joseph Y-T., and Jennifer Whitehead. "On the complexity of fixed-priority scheduling of periodic, real-time tasks." *Performance Evaluation* 2.4 (1982): 237–250.

Liu, Jane W.S. *Real-Time Systems*. Upper Saddle River, NJ. Prentice-Hall. 2000.

Li, Qing, and Caroline Yao. *Real-Time Concepts for Embedded Systems*. CRC Press, 2003.

Liu, Chung Laung, and James W. Layland. "Scheduling algorithms for multiprogramming in a hard-real-time environment." *Journal of the ACM (JACM)* 20.1 (1973): 46–61.

Lutz, Robyn R. "Analyzing software requirements errors in safety-critical, embedded systems." *Requirements Engineering*, 1993., Proceedings of IEEE International Symposium on. IEEE, 1993.

Mok, Aloysius K. "Fundamental design problems of distributed systems for the hard-real-time environment," *Technical Report*, Massachusetts Institute of Technology Cambridge, MA, USA (1983).

Mok, Aloysius K., and Deji Chen. "A multiframe model for real-time tasks." *IEEE Transactions on Software Engineering* 23.10 (1997): 635–645.

Nilsson, Johan, Bo Bernhardsson, and Björn Wittenmark. "Stochastic analysis and control of real-time systems with random time delays." *Automatica* 34.1 (1998): 57–64.

Noergaard, Tammy. *Embedded systems architecture: a comprehensive guide for engineers and programmers*. Newnes, 2012.

Rajkumar, Ragunathan, Lui Sha, and John P. Lehoczky. "Real-time Synchronization Protocols for Multiprocessors." *RTSS*. Vol. 88. 1988.

Rajkumar, Ragunathan. *Synchronization in real-time systems: a priority inheritance approach*. Vol. 151. Springer Science & Business Media, 2012.

Ramamritham, Krithi, and John A. Stankovic. "Scheduling algorithms and operating systems support for real time systems." *Proceedings of the IEEE* 82.1 (1994): 55–67.

Seto, Danbing, et al. "On task schedulability in real-time control systems." *Real-Time Systems Symposium*, 1996., 17th IEEE. IEEE, 1996.

Sha, Lui, Ragunathan Rajkumar, and John P. Lehoczky. "Priority inheritance protocols: An approach to real-time synchronization." *IEEE Transactions on Computers* 39.9 (1990): 1175–1185.

Sha, Lui, et al. "Real- time scheduling theory: A historical perspective." *Real-Time Systems* 28.2–3 (2004): 101–155.

Sprunt, Brinkley, John Lehoczky, and Lui Sha. "Exploiting unused periodic time for aperiodic service using the extended priority exchange algorithm." *Real-Time Systems Symposium*, 1988, Proceedings. IEEE, 1988.

Sprunt, Brinkley, Lui Sha, and John Lehoczky. "Aperiodic task scheduling for hard-real-time systems." *Real-Time Systems* 1.1 (1989): 27–60.

Spuri, Marco, and Giorgio Buttazzo. "Scheduling aperiodic tasks in dynamic priority systems." *Real-Time Systems* 10.2 (1996): 179–210.

Stankovic, John A., et al. *Deadline scheduling for real-time systems: EDF and related algorithms*. Vol. 460. Springer Science & Business Media, 2012.

Strosnider, Jay K., John P. Lehoczky, and Lui Sha. "The deferrable server algorithm for enhanced aperiodic responsiveness in hard real-time environments." *IEEE Transactions on Compute*rs 44.1 (1995): 73–91.

Stroustrup, Bjarne. *The C++ programming language*. Pearson Education, 2013.

Tindell, Ken W., Alan Burns, and Andy J. Wellings. "An extendible approach for analyzing fixed priority hard real-time tasks." *Real-Time Systems* 6.2 (1994): 133–151.

Weiss, Mark. *Data Structures & Algorithm Analysis in C++*: 4th edition. Pearson Education. 2013.

Xu, Jia, and David Lorge Parnas. "Scheduling processes with release times, deadlines, precedence and exclusion relations." *IEEE Transactions on Software Engineering* 16.3 (1990): 360–369.

Yiu, Joseph. *The Definitive Guide to ARM® Cortex®-M3 and Cortex®-M4 Processors*. Newnes, 2013.

INDEX

Chapter 1

Chapter 2

Chapter 3

Chapter 4

Chapter 5

Chapter 6

Chapter 7

9 781516 501885

Printed by Libri Plureos GmbH in Hamburg,
Germany